THE CONTRASUMERS

THE CONTRASUMERS

A Citizen's Guide to Resource Conservation

Albert J. Fritsch

PRAEGER PUBLISHERS
New York • Washington

Published in the United States of America in 1974
by Praeger Publishers, Inc.
111 Fourth Avenue, New York, N.Y. 10003

© 1974 by Praeger Publishers, Inc.

All rights reserved

Library of Congress Cataloging in Publication Data

Fritsch, Albert J
  The contrasumers.

  1. Energy conservation. 2. Power resources.
3. Environmental protection. I. Title.
TJ163.3.F74   333.7   74–6726
ISBN 0–275–10140–1
ISBN 0–275–63540–6 (pbk.)

Printed in the United States of America

Contents

List of Tables	vi
List of Figures	vi
Preface	vii
1 The Second Hundred Years	3
2 Consumerism Versus Environmentalism	20
3 Comprehending the Crisis	37
4 Assessing Our Needs	65
5 Conservation Strategies	96
6 A Radical Environmentalism	133
7 The Next Hundred Years	155
Lifestyle Index	159
Sources for the Lifestyle Index	180

List of Tables

I. Energy Consumption in the United States, 1876 and 1976 — 18
II. Number of Motor Vehicles in the World, 1971 and 1972 — 26
III. Energy Consumption, Developed Countries, 1971 — 27
IV. U.S. Gross Consumption of Energy Resources, by Sources and Consuming Sectors, 1972–73 — 40
V. U.S. Coal Estimates, 1970–85 — 45
VI. Costs of Pollution Damage in the United States, 1968 — 60
VII. Electric Appliances in U.S. Homes, 1972 — 100
VIII. Estimated Power Consumed by Electric Home Appliances in a Year — 100
IX. Significant End Uses of Energy in the United States — 131
X. Areas of Major U.S. Potential Energy Savings — 132
XI. Annual Energy Units per Capita in Selected Countries — 178

List of Figures

1. U.S. Gross and Net Energy Consumption, 1947–2000 — 29
2. U.S. Gross and Net Energy Inputs per Capita, 1947–2000 — 29
3. World Geographic Areas in Proportion to Oil Reserves — 43
4. The "Fossil Candle" — 65
5. U.S. Energy Consumption by Sector, 1971–2000 — 128
6. U.S. Energy Consumption by Energy Source, 1971–2000 — 128

Preface

For almost two hundred years Americans have taken for granted an ever rising standard of living. We have evolved into the greatest consumer culture in history, constantly setting new world records for energy consumed, cars owned, and beer cans thrown away. While these consumption habits are not exclusively American (overconsumption knows no racial, social, political, or geographic boundaries), the United States has led the way in spreading the doctrine of commercialism throughout the world. Canada, Western Europe, Australia, Japan, and now Russia and Eastern Europe are rapidly catching up.

Every plant and animal consumes material things in order to grow, but plants and animals thrive in a balanced ecological system where a certain fundamental order and harmony prevail. Man has introduced disorder into this system. His uncontrolled consumption of resources has taken its toll of the world environment. Chemical pollutants, such as lead, mercury, and cadmium, not to mention the thousands of chemicals used as pesticides and plasticizers, are cropping up in the tissues and vital organs of men and animals in the most distant and remote parts of the globe.

On the credit side, the United States has begun to witness an emerging counterculture that advocates unorthodox conservation practices. The American Way is coming under fire from within, and our very lifestyles are being questioned by futurists, organic food faddists, bikers, and citizens living in communes. But conservation in the midst of apparent plenty is hard for many Americans to stomach. The egocentric consumer who values his gasoline, his beef, and his electricity takes the attitude that basic changes will occur only in the shadow of such disasters as severe energy and mineral shortages, a crippling dollar drain, worldwide famine, or acute environmental pollution.

This country has had its share of rude awakenings since the century began. World War I did not make the world safe for democracy. World War II did not rid the world of dictators. Sheer physical prowess did not win out in Vietnam, and the people of the Third World no longer look to us as exemplars of progress. The latest rude awakening has been brought about by blackouts, brownouts, fuel shortages, traffic congestion, air pollution, rising food prices, and a U.S military budget that hovers around $100 billion.

Meanwhile, the rest of the world clamors for a slice of the limited resource cake. With limits to material growth becoming more and more evident, the inescapable conclusion is that our country's mature destiny must be global and social in character and qualitative rather than quantitative in scope. It is the purpose of this book to point out how such a changeover can be effected without producing major disruptions in the U.S. economy. In short, *The Contrasumers* is intended to be both a blueprint for a service-oriented economy, wherein Americans can relate in a meaningful manner to other peoples of the world, and an exercise in old-fashioned citizenship.

The Lifestyle Index at the back of the book will enable the more venturesome reader to evaluate his or her own lifestyle. It is sincerely hoped that filling in the blanks and adding up the totals in the Index will turn a number of overconsumers into true *contrasumers*.

A book such as this touches upon many fields, from ecology and environmental law to economics and ethics, and assistance was necessary in assembling the materials. Fortunately, the Center for Science in the Public Interest in Washington, D.C., has a number of dedicated staff members and advisers whose help proved invaluable.

I owe special thanks to our staff environmental engineer, Barry Castleman, who gathered much of the energy data used in the Lifestyle Index. I also am grateful to Jack Egan and Ralph Gitomer for proofreading and contributing to the energy sections; to student interns Nick Lefevre, for environmental law history, and Dan Price, for electric equity rates; to our research associates Barbara Hogan, for compiling data on American life in 1876, and Bill Millerd, for suggestions on nuclear power; and to my co-directors at the Center for Science in the Public Interest, Mike Jacobson and Jim Sullivan, for constructive criticism during the course of this work.

Among the volunteers at the Center, I want to thank Sherry Carstenson, for typing; Carter Lee, for technical assistance; Art Purcell, for resource recycling strategies; Debbie Purcell, for helping to update statistics; JoAnn Scott, for sections dealing with home economics and household conservation; and John Waring, for the calculations appearing in Table I. I am also grateful to Bob Duggan and Pete Henriot for valuable criticism of the ethical principles and the international strategies presented in Chapter 4.

Thanks are due several U.S. Government officials who helped with the statistics used in the Lifestyle Index: Eileen Claussen and Larry Deister of the Environmental Protection Agency, Russ Wylie of the Postal Service, John Gale of the Department of Agriculture, Eric Hirst of the Federal Energy Office, James West of the Department of the Interior, and Wayne McCaughey of the Bureau of the Census. I am most grateful to the following also: Matt Reilly of the National Academy of Engineering, Charles Fritsch of Bell Laboratories, Robert Griffith of the American Gas Association, and Robert Herendeen of the Center for Advanced Computation of the University of Illinois.

THE CONTRASUMERS

1 · The Second Hundred Years

Americans like birthday parties. July Fourth 1976 will mark our two hundredth year as an independent people. In preparation for this noble anniversary, let's take ourselves back to the first American Centennial celebration and then look ahead to the next one.

In 1876 we had just begun to recover from the most costly war in our history. We were beginning to stir, to forget past hostilities, and to expand across the West. We had our Indian wars, a major depression, and some of the worst corruption in our nation's history, but somehow all these troubles did not dampen the hope and dreams of the hordes of immigrants coming to our shores. Ours was a land of raw vitality and constructive disarray.

Now let's look into the future. The year is 1984, the time of Big Brother and *Brave New World*. What if that year finds us with the same rate of consumption and the ever increasing rate of commercialism that we experienced in the early 1970s? Visualize vast tracts of corporate farms, largely devoid of people, dotted with abandoned farmhouses. Think of sprawling suburban strips and monolithic apartment buildings surrounding the hollow shells of cities linked by networks of highways. See a land being gobbled up at the rate of 200,000 acres a year by power shovels stripping coal to fuel the plants that produce our electricity.

Imagine superhighways clogged with 200-plus–horsepower cars that last for less than ten years and then are junked for new ones—all competing for a narrow section of concrete, consuming precious gasoline, and forcing their emissions through delicate catalytic devices that always need repair. Frustrated, afraid, self-centered, nervous, disgruntled over rising prices and increasing crime, the American of 1984 moves on to an uncertain future.

Whether we accept Huxley's "brave new world" or one that seems more probable from our present perspective, we find them equally frightening. Both accept the American Way, with its production records, increasing GNPs, and drive for material conveniences—bizarre scenes of people totally taken up with their own cares and with little regard for what the rest of the world is doing. The visions of the future do have a continuity, however, with the character and dreams of Americans of another century.

Nineteenth-century America was confident in the power of its new inventions: the tractor, the steam warship, and the telephone. It was convinced that the world had crossed the barriers of the unknown. It placed all trust in the Machine and the Gadget, considering them gods. It knew little about the tensions and stresses of a later century that would again seek nonmechanical ways of influencing human progress.

Our century is certainly different from the nineteenth, and each decade of the twentieth century is different from the previous one. We have airplanes, autos, computers, and nuclear power. We also have air raids, congestion, invasion of privacy, and radioactive pollution. World Wars I and II, the Korean War, and the Vietnam War are now part of our historical memory, but they have all left profound marks on the American character. We are becoming more subdued, less bombastic, more cautious, and less isolationist. We are the fruition of that other century's dreams.

The Machine and Gadget have become part of the American Way; it is patriotic to build more of them and to build them better. They have become part of our lifestyle; they dictate where we go and what we enjoy. They are the foundation stones of a consumer culture based on mobility, convenience, and fashion. Being mobile, Americans aren't going to stay in one place for long. Lovers of convenience, they are not going to overromanticize their rustic past. As men and women of fashion, they live for the moment and reflect little about the future.

To compare our American past with the contemporary scene,

we must make some basic assumptions: first, that the "average" American represents a goodly number of citizens. When we divide a quantity of consumer goods by a population to obtain per capita figures, we are able to arrive at some notion of what a collective group of consuming individuals does. But who is this "average American"? Does he or did he ever exist? We can try to find out by taking the total population and subtracting out various minority groups: Indians and blacks, the young and the aged, Catholics and Jews, rural people, the brown, the red, and the yellow, the underemployed and unemployed and self-employed, college graduates and college students, teachers and doctors, and finally the largest minority, the American male (there were 4.7 million more American females in 1972 than males).

After eliminating all of these groups we are left with a middle-aged, Anglo-Saxon, Protestant, native, urban, white female who didn't graduate from college. There are only about 10 million of these "average" Americans. They constitute a far smaller group than other, better-publicized minorities.

Americans are unique in the way they think, speak, act, and treat one another. When we walk down the streets of London or Paris, the English or the French know we are Americans, even before we open our mouth or wave a credit card. Our uniqueness is connected with our consumption habits, the way we spend, exploit, and discard material things. We are captivated by telephones and superhighways; we drive powerful and bulky cars; we supercool or superheat our homes; we drink hot coffee with cold ice cream.

Consumption is America's obsession. Over 40 per cent of the world's aluminum and manganese and a third of the world's fuel are consumed in this country. We own almost half the world's passenger cars. We lead the world in pollution—the waste products of consumption—spilling off almost two tons of waste per person per year. We did not arrive at this stage overnight. Burning forests proved to be a rapid way to make farmland for our pioneer forebears. Shooting buffalo to clear the railroad rights-of-way was great sport.

Are Americans really so different? Let's look at two "average" families, one rural, the other urban but both white, middle-class, and English-speaking, with school-age children.

The Rural Setting

The sun rises over a small hamlet on the east fork of the Little Miami River in Ohio, some fourteen miles from Cincinnati. On this day the center of America's population passes through Perrintown and the home of John and Mary Davis. While their farm does not share in the flatness of the Great Plains, which begin a few miles away and stretch as far west as Denver, it does have a rolling contour that is more typical of America's farmland.

The Davis farm is still partly in debt, but the land is fertile and deep-soiled, and there is a large stand of black walnut trees, a sure sign of good land. Farm prices could be higher, especially with present economic conditions, but John and Mary are voting Republicans, like most of their neighbors. They don't care much for Southern Democrats. The Davises are hardworking, churchgoing, frugal folk who like to wave the flag and eat apple pie.

Rural America appears sleepy in midsummer, but on closer inspection wheels are spinning and people are bustling about. The season of dawn-to-dusk labor is at hand, for this is Monday, July 3, the day before the most American of days. A little more time spent in preparation today makes the fireworks, parades, and fried chicken better appreciated tomorrow.

Under his weather-beaten straw hat and behind a layer of dust and sweat stands John Davis, American independent producer par excellence: a middle-aged army veteran, adequately but not excessively educated, possessor of common sense and a variety of skills ranging from carpentry and bookkeeping to dairying and gardening. He is a jack-of-all-trades and fair master of many. He rises with the sun and gets tired soon after it sets. Artificial light hurts his eyes, so he never reads more than the local newspaper. He's not much for night life but does enjoy a day in town, a chat after church, and an occasional barn-raising, fire, election, or family reunion.

John is his own weatherman, giving only limited credence to the views of forecasters and the yearly almanac. His habit is to go outside after rising and sense the air, humidity, cloud cover, and light in the sky. One could dismiss this as intuition, but John knows when it is going to rain, how soon it will begin, and when it will be over. It's one of his agricultural skills.

The Davis farm lies between Dixie and the Great Plains. Mixed farming is the custom, and John raises hay, corn, wheat, barley, and even a small patch of tobacco for local consumption. He has cows for milk, chickens for eggs, hogs for meat, a garden for vegetables, and an orchard for apples, plums, pears, and cherries. These require a rich variety of seasonal tasks that offer spice to an otherwise dull life—land-clearing and rock-hauling, plowing and cultivating, planting and harvesting, wood-cutting and fence-mending.

Life on the Davis farm is occupied with concerns and worries unknown to city dwellers. The rye is good this year, the wheat average, and the pears poor. The corn is knee-high by the Fourth of July. The clover has been injured by rain, but the rest of the hay is tops.

The farm day starts with a sizable breakfast: hominy with cream and maple syrup, occasionally beefsteak with creamed potatoes or buckwheat cakes, or maybe eggs and bacon with apple pandowdy, sweet sauce, and coffee. A good cook and plentiful food fortify a farm family for the rugged haying season. A full breakfast is remedy for the sharp aches in John's back—reminders of army life in the Carolina swamps. A good breakfast beats all the drugs and exotic liniments one can buy.

This farm day is both typical and unique. Hay has been loaded and stacked. The hired hands have worked hard and fast. The day is punctuated with little things—a slip and fall due to slick shoe leather from walking in hay, a bent pitchfork prong in need of straightening, a black snake that gets away, the temperament of animals bothered by horseflies, a moment of concern over weeds in the corn, a water jug run dry, goose-pimples caused by a sudden cool breeze on sweaty skin.

Mary's unliberated day is a bit longer than John's. She's content in her workaday world. Mary prepares three meals a day for a hungry family. She washes with few conveniences, mends a mountain of clothes, goes to the store but once a week, tends a flock of chickens, snatches an hour here or there for flowers and garden, keeps a tidy house, and even has an extra day or so to help when farm work gets heavy. While she sends the children to the local school, she still feels it is her duty to put in some time teaching them at home.

There are several unpleasant features to Mary's day. Although the water is cool and soft, it has to be carried. Although eggs are fresh, chickens need feeding. Although creamy butter is enjoyed by all, it has to be made by hand. The house is Mary's domain, and a needed repair here or a coat of paint there has to be detected and remedied. It is Mary's job to sweeten the smelly outbuildings with lime. She sees that weeds are removed from the potato and cabbage rows, that tomato stakes are in place and berries picked. Dinner, a midday affair in rural America, has to be cooked for four sweaty men. That means amassing a meal of garden peas, homemade bread, smoked ham and fried chicken, pickles, lettuce with fried bacon and vinegar dressing, boiled potatoes, and some rhubarb pie.

Mary's skills exceed even those of her husband. She knows how to kill a chicken or skin a rabbit, how to pick such herbs as rosemary, parsley, mint, and dill, how to pickle a host of vegetables, make homemade preserves, gather seasonal nuts and berries, choose greens and mushrooms, smoke meat in winter, and make ice cream in summer. She is talented at baking cookies and blackberry cobbler. She can squeeze apple cider and prepare vinegar. While she doesn't know how to spin—that art died with her grandmother—she can still select a bolt of cloth and make her own clothes and some of the children's.

On this day there are the added tasks of preparing for the upcoming picnic. Two young pullets are caught, decapitated, scalded, picked, and dressed. Mary stokes up the range for an extra batch of rolls and goodies; stuffs eggs and tops them with a dash of parsley; selects some red plums and June apples that are not too wormy; and puts the perishables in the springhouse for tomorrow. Nothing is forgotten.

Summer is lush but demanding. The long summer days seem remote from the mud, snow, and cold of winter, from herb tea, rosy fireplaces, quilt-making, and shelling nuts for the Christmas cake. But summer at the Davis house can also be unpleasant. There is no such thing as air conditioning on hot humid nights; flies buzz constantly; frequent thunderstorms bring threats of hail and lightning. Barn smells float houseward when least wanted, and horses' hooves raise a dust storm on the macadamized road in front of the house.

Before turning our attention to urban America, let's add an important detail to the above scene. John and Mary Davis are preparing for July 4, *1876*. And in 1876 there are no automobiles, electricity, central heating, indoor plumbing, telephones, refrigerators, radios, or TVs. There are only buggies, kerosene lamps, fireplaces, outhouses, springhouses, and wood ranges. The only tractor John ever heard of is a lumbering steam engine that creeps in for threshing each August. The Davis world is devoid of power lines, superhighways, and aluminum cans. It thinks itself bustling when it recalls the hardships of earlier frontier days, but its pace is snail-like by twentieth-century standards.

The Urban Setting

The ring of the anvil at the local blacksmith is no more; it has been replaced by the rush of cars. The center of U.S. population has moved steadily westward across Indiana and Illinois. By July 4, 1976, it has reached a suburb somewhere in the metropolitan area of St. Louis. The household of Bert and Barbara Kelly includes two children, two cars, two pets, two baths, and a $30,000 mortgage. Bert is an engineer employed at Rockwell, fifteen miles and forty minutes away from his home via the beltway.

Twentieth-century America is urbanized, electrified, and motorized. Nine out of ten people live in towns and cities, whereas most of the population lived on farms in 1876.* Electricity is the Kellys' lifeblood. From the moment they wake up to the electric alarm, they enjoy its mysterious power and energy. Lights go on, razors buzz, toothbrushes vibrate, hair dryers groom. Then come the coffee percolator, toaster, electric skillet, and later the washer and dryer for last-minute laundry before the bicentennial celebration. The radio brings the morning news from around the world, and

---

* In the 1870s less than a third of the population of the United States lived in cities of 2,500 or more. During that time, however, agricultural workers were slowly being outnumbered. In 1870, 6,850,000 of the employed worked on farms, and 6,075,000 did not. By 1880 farm workers numbered 8,585,000, but a larger number of workers (8,807,000) were employed in nonagricultural jobs.

TV placates the kids for a while. Time is conserved through electricity; what would take Mary and John Davis a morning of hard work now takes less than an hour.

Day and night the steady hum of an air conditioner keeps the Kelly house cool in the sweltering July weather. The electric water heater and dishwasher work overtime. The deep freeze adds to convenience. The refrigerator keeps eggs, milk, and vegetables fresh. Before the day is over, the electric can opener, the disposal, and the kitchen range will have seen lots of use. The amount of energy consumed this day by the Kellys will be ten times that consumed by the Davis family a century ago.

The Kelly family is motorized. It doesn't feed and groom its means of locomotion or spend a week or more each year putting in hay to feed the horses. However, a large part of Bert's $25,000 yearly earnings goes for auto expenses on the two cars—about $900 for depreciation, $400 for repairs and maintenance, $800 for gasoline and oil, and another $400 for insurance, licenses, tires, and accessories. This 10 per cent of the family budget does not include auto taxes, tolls, and parking fees. Bert has to work about five weeks a year just to stay motorized.

On this particular Saturday the Kelly household is on the move. Bert must rise early to get to the golf course on time for the customary weekly sporting event with his three friends. He is no midweek golfer, and the sacredness of this once-a-week event is important to him. Barbara drives young Mary to her weekly clarinet lessons in the station wagon and then returns to make some morning phone calls about her next garden club meeting. Fourteen-year-old Greg rides his trusty bike over his delivery route and gets home early for brunch before racing off to spend the rest of the day at the local swimming pool.

On the way home, Bert stops for gas. Finally, it is midafternoon and time to prepare for an evening cookout for some old friends. After a shower, Bert catches a few moments in an easy chair, just sitting and watching the final holes of the golf playoffs on TV. Barbara doesn't indulge in such luxuries; she has to assemble some items for the picnic tomorrow (hardly a bother since she uses mostly prepared foods, purchased at the supermarket). While she prepares a large salad for the evening's cookout, she remembers to prompt her husband—the pyrotechnician—to get the charcoal hot for the hamburgers.

The smell of cooking food greets the kids coming home from the swimming pool, wet, stringy-haired, and hungry. Barbara feeds them early so they can go to the park with friends to watch the fireworks after sunset. Eventually, the guests arrive and sit for a while and socialize over gin and tonic. Perhaps no other meal of the week is eaten by Bert and Barbara at such a leisurely pace.

This Saturday has been no more frantic than any other average summer or winter Saturday. All the Kellys take their hobbies, sports, and social life seriously. Few of these are really family events, for each member has his or her particular tastes. The family is respectful of individual interests, but there is the usual kidding and joking about success and failure on the links, at the flower show, at the music festival, or in the swimming match. Their superactivism makes one wonder if they could sit through an entire old-fashioned picnic.

The Kellys' lives revolve around the superhighway, which takes the family members to and from school, church, club meetings, and work. Time, not distance, is the basic unit of motorized measure. In August the Kellys plan to go to Maine, which is more than two but not quite three days away, via a 1,200-mile ribbon of concrete, unbroken except for a monotonous series of exit markers indicating places with funny-sounding names.

Roads fitted into the life of the Davis family in the nineteenth century in a somewhat different way. Travel was slower, and every turn in the rural lanes offered a scene worth observing. But roads then also meant washed-out bridges, dust and mud, and the clip-clop of horses. Perhaps the Davis family of the 1870s sensed the achievements of rapid transportation more than their counterparts of this century. Their own fathers and mothers had made the pioneer trip west on canal boat, log river raft, and covered wagon and considered it the experience of a lifetime. Puffing iron horses had cut to a fraction the travel time from the Atlantic to the Ohio Valley at the first Centennial. And in the last third of the nineteenth century, black soot and racketing rails were signs of progress, not types of pollution. A trip to Philadelphia for the special Ohio Day at the Centennial in 1876 took a few days by train—and just a little longer than it would take today on Amtrak!

The Good Old Days?

Yesterday's luxuries are today's necessities, but they can be a mixed blessing. Electric appliances are expensive to repair. Modern food is bland and filled with chemical additives. Airplanes disgorge passengers miles from town. Highways are crowded and pollution-ridden, and driving is hazardous. Human relationships seem to have been sacrificed in the name of speed and convenience.

Old-timers remember the sight of children following the ice wagon, the gliding ride of the streetcar, the taste of homecured hams, the sound of the clear stream under the creaky covered bridge. They tell of old Dobbin's ability to find his way home after a party while the driver takes a boozy snooze. Sunflowers, bonnets, scythes, and cowbells seem remote, but they strike familiar chords in modern Americans, who are often only two generations removed from the farm. In fact, a Gallup poll of May 12, 1973, showed that 56 per cent of Americans would prefer a rural life if they could have it and 25 per cent a suburban life. Only 18 per cent wanted to spend their lives in the city. It seems paradoxical, but one of the biggest differences between today and a century ago is that more Americans today want a rural life, although fewer of them know what a rural life is really like.

Before allowing ourselves to be lulled by memories of the "good old days," let's recall what life was really like a hundred years ago. The 1870s was a decade of immense construction activity and mobility. Over 2,500 miles of railroad track a year were being laid, mostly by the muscle and sweat of immigrants just off the boat from Europe. The steam-driven reaper, one of the great inventions of the century, was in the process of development. At the Philadelphia Exposition of 1876, an obscure inventor by the name of Alexander G. Bell demonstrated an invention that caught the fancy of the Emperor of Brazil and received the publicity that made telephone a household word in the short space of a decade.

The year 1876 also marked the midpoint of a great depression, a period when imports of silks dropped a third, of tea a half.

Wages fell 15 per cent, and charges for professional services were halved. In June, a period of frequent fights between trainsmen and bums, two hundred tramps stole a Rock Island railroad train. Business failures, which had numbered 7,740 in 1875, rose to 9,092 in 1876. The labor movement was beginning to surface, but businessmen continued to oppose every effort by workingmen to improve their economic lot.

Post-Civil War chaos led to rampant vice in the cities. Children picked pockets in the streets. Girls sold matches, flowers, and hot corn—a "cover" for selling their bodies. Adequate water supplies were practically nonexistent. There were no decent public health programs or building codes. Police were ill trained, and most government employees were political hacks. Blacks, one-tenth of the total population, were "held in place," excluded from streetcars, theaters, and restaurants. Jews were not allowed in the Cataract House at Niagara Falls. Catholics were frowned on as poor ignorant foreigners, and Indians were mere savages, clinging to the fringes of civilization. Anglo-American dominance was taken for granted. The majority of the people had to be satisfied with dirt, sweatshops, unpaved streets, and life on isolated farms.

If any decade should not have been romanticized, it was the 1870s, but people living at that time did have strong hopes for better things. There was a sense of community even in the urban ghettos. There was time to talk and laugh. There was a human quality to life and to the way people faced the fact of death. The rural lands were more thickly and evenly inhabited than today. In the small towns newly built homes had running water and gaslight. Central heating, while not widespread, was the rule in the homes of the wealthy. Kitchen fireplaces were being bricked up, replaced by wood-burning ranges. In affluent homes, the furniture was elaborately decorated. The typical parlor had a portiere hanging across a doorway, a whatnot shelf for bric-a-brac, daguerreotypes in folding covers, wax flowers under glass domes.

Faster ships and refrigerated railroad cars were beginning to bring a variety of foodstuffs to urban tables. Strawberries were available for four months of every year, grapes and peaches for six, and sweet corn for nine in the larger cities. Cluttered country stores in rural areas offered smoked meats, flour, and meal, but little by way of variety except in local produce at midsummer.

Carbonated soft drinks, a major supermarket item today, were curiosities at the Philadelphia Exposition of 1876.

Lifestyles—Then and Now

Lifestyles that make use of the most sophisticated technological advances do not necessarily afford a greater enjoyment of life. Current technology, for example, has taught people to sacrifice the taste of sun-ripened food in favor of prepared foods treated with food coloring, artificial flavoring, and preservatives.

Any comparison of lifestyles of different eras is bound to be difficult and perhaps misleading. While material possessions can easily be measured, such feelings as joy, friendship, and a sense of freedom or community cannot. Contemporary social statistics for mental illness, alcoholism, church attendance, prevalence of serious crime, or frequency of suicide are largely lacking for earlier centuries. Nevertheless, certain differences can be pointed out if we make allowances for the word "average" as applied to citizens in a pluralistic society. The following comparisons focus on the lifestyles of average Americans of the 1870s and the 1970s.

1870s	1970s
HOUSEHOLDS	
Two-story frame structure. Poorly manicured yard but well-tended orchard and garden. Dusty road in front of house. In back, chicken house, outdoor toilet, coal- or woodshed, springhouse and smokehouse, clotheslines, fireplace for heating wash water. Indoor plumbing in some urban homes only. Kerosene lamps or gaslight.	Apartment or ranch-type brick building. Little or no yard, shrubs well trimmed. Cement sidewalks, paved streets with storm sewers and street lighting. Garage or carport. Utility room with automatic washer and dryer, furnace, water heater, and storage space. Indoor plumbing, central heating, electric lighting, and air conditioning.
Large airy kitchen with wooden cabinets, churn, wooden chairs and table, and coal or wood range. Dark living room with over-	Small kitchen with artificial lighting and electric stove, toaster, can opener, refrigerator, and dishwasher. Large living room

1870s

stuffed chairs, sofa, figured wallpaper, family photos, piano or organ, china cabinet, and cuspidor. Master bedroom with large bed, washbasin and pitcher, heavy dresser, quilts and feather mattress.

1970s

with television, easy chairs, record player, telephone, wall-to-wall carpeting. Air-conditioned bedrooms. Spring mattresses. Adjoining bathrooms with showers.

WORKING CONDITIONS

Majority of workers in agriculture, small crafts, and construction. Minority in heavy industry and the professions. Distance to work about one mile. Twelve-hour workday, with main meal at noon. Wages $1.00 to $2.00 a day. Few vacations. Child labor, no unions, few employed women, and little unemployment.

Majority of employed persons urbanized and service-oriented. Businessmen and professionals well paid. Minority in heavy industry and agriculture. Average distance to work about fifteen miles. Eight-hour workday. Average income $8,000 to $15,000. Paid vacations, insurance, and retirement benefits. Women 40 per cent of labor force. Unemployment 5 per cent.

EDUCATION

Half of children in the South and most in the North in primary schools (average duration four–six years). College for rich; university life barely beginning.

Universal education for children. College attended by half of eighteen–twenty-one-year-olds. Private schools expensive, community colleges growing rapidly.

DIET

Seasonal and regional foods, prepared with spices and herbs. Smoked and cured meat, home-grown, sun-ripened produce. Home baking, hand-churned butter, fresh-ground coffee, draft beer, unrefined sugar products. Food purchases 10 per cent of income. Good cooking practices,

Large variety of fruits and vegetables. Fresh meat with moderate seasoning, chemical additives, and preservatives. Bakery bread, freeze-dried coffee, margarine, bottled and canned soft drinks and beer, refined sugar, TV dinners, frozen foods, cold cereals, cake mixes. Food purchases 23 per

1870s

flavorful foods, balanced diet in summer and fall but not in winter.

1970s

cent of budget. Loss of cooking art due to convenience foods. Poor nutrition widespread.

HEALTH

Life span less than fifty years. Poor hospitals and public health facilities. Patent medicines, fake cures, midwives. Personalized medical care, home visits from physicians. Poor dental care, susceptibility to mouth diseases among poor. Aged members cared for at home. Few welfare institutions. Funerals held in home with burial in churchyard or on farm.

Life span more than seventy years. Good hospitals, high hospital bills. Few home visits, less personal care. Regulated drugs, across-counter sales. Expensive dental care, with most tooth decay due to sugar diet. Old-age homes. Social opportunities for handicapped and retarded. Commercialized funerals with burial in high-priced cemetery plots.

RECREATION AND CULTURE

Weekly visits to relatives, family reunions. Church gatherings, barn-raisings, picnics and outings. Fishing in summer, sledding in winter. Trips over a day's journey very rare. Beginnings of tourism, visits to beach resorts, mineral baths.

Television, radio, stereo. Cookouts, cocktail parties, drive-in movies. Summer boating and swimming, winter ski trips, football. Extensive vacation travel, airplane trips, conventions, tours to Europe. Evening drives in the country.

Power and Energy—Then and Now

Few things distinguish nineteenth- from twentieth-century man more than the difference in the amounts of power at man's disposal. For nineteenth-century man, power meant the ability to move a heavy load.* It meant champion Belgian workhorses at

---

* Horsepower is only one of the units used for measuring power (the time rate for doing work). Power units also include watts (for electricity) and foot-pounds per second. One horsepower is the equivalent of 746 watts, or about 550 foot-pounds per second.

the county fair pulling sleds piled high with oversize loads of wheat or rock. Horsepower was more than a picturesque relic of power measurement in the nineteenth century. In the first census year following the Centennial, work animals furnished more than 40 per cent of the total horsepower in this muscular land. The untiring iron horses of the American railroads came next with about 33 per cent:

Source	Horsepower
Work animals	11,580,000
Railroads	8,592,000
Factories	3,664,000
Mines	715,000
Merchant ships	714,000
Farm steam engines	668,000
Windmills	40,000
Other	314,000

Altogether, 26,314,000 horsepower were produced and consumed in the United States in 1880, or slightly more than one-half horsepower per person.

While twentieth-century man lacks a feeling for "true horsepower," he does know the surge of an auto engine. Behind the wheel, he commands power at his finger- and toetips; he plunges his 2-ton automated monster from rest to sixty miles an hour in mere seconds. Power gives him an exhilarated feeling of commanding his own destiny. He is mobile; he moves fast to new places. He degrimes and shines his 200-horsepower idol with religious fervor. When one of America's 110 million drivers gets behind the wheel, he is driving the equivalent of a 100-team chariot. Accustomed to a feeling of power never sensed by a Roman general, how can he be affected psychologically?

While we are familiar with the surge of a powerful motor, most of us are barely aware of the phenomenal growth of power in the United States in the past thirty years. Most of this increase can be traced to our cars. In 1940 the average horsepower per American was twenty-one. This climbed to thirty-four in 1950, to sixty-two in 1960, to eighty-one in 1966, and to ninety-three in 1971. Moon rockets, airplanes, and industrial machinery were not responsible for those sharp increases. Some 95 per cent of U.S. horsepower is in motor vehicles.

Americans are more power-sated than power-hungry. We want things done in a hurry at great speed and with the least human effort. We talk and think in terms of physical power. A few bombs dropped at the right time and place will soften the minds of the most obdurate enemy. A Marine landing will correct most errors of foreign policy; a loud siren will disperse a mob; a powerful street light will deter crime. We have become a people of power politics, power brakes, and power plays.

The concentration of physical power in our cars is wasteful, for it requires a major portion of our natural resources just to keep them poised for action. Petroleum, a nonrenewable resource, provides most of our current energy, whereas a century ago two-thirds of our energy was from renewable wood and waterpower sources (Table I). Admittedly, burning wood was a terribly inefficient source of energy. This might account for our per capita energy consumption merely quadrupling while power production increased two hundredfold. However, this increase has resulted in a terrible drain on our store of nonrenewable energy.

TABLE I
ENERGY CONSUMPTION IN THE UNITED STATES, 1876 AND 1976

	1876	1976
Total population	46,107,000	216,770,000
Energy (in trillion Btu's)		
from mineral fuels	1,600	80,000
from waterpower	182	3,650
from fuel wood	2,868	275
Per capita (in Btu's per year)	100,860,000	387,161,500

SOURCES: Figures are one-year straight-line projections of official government figures for 1875 and 1975. National Coal Association; U.S. Bureau of Mines; U.S. Forest Service (fuel wood data); U.S. Bureau of the Census (midpoint of "D" and "E" projections for 1976); U.S. Department of the Interior (W. G. Dupree, Jr., and James A. West, "United States Energy Through the Year 2000," December 1972).

The nineteenth was the last century of the horse, and its replacement by the machine was accomplished with great reluctance.

As late as World War I the British attempted a cavalry charge or two. Horses are sensitive and respond to human urging. The machine is powerful and speedy but dumb. Horses deposit organic wastes, which scented the streets of every nineteenth-century town —but made roses grow. Horses need stables in winter and pastures in summer. For their part, cars need garages, highways, parking spaces—and graveyards. They also emit organic wastes, but auto emissions, which account for over 40 per cent of man-made pollutants, have never helped a rose to grow.

Auto enthusiasts are quick to point out the shortcomings of the horse age, but they were outdone in 1899 by the *Scientific American,* which had this to say of the coming auto age:

> The improvement in city conditions by the general adoption of the motor car can hardly be overestimated. Streets—clean, dustless, and odorless—with light rubber-tired vehicles moving swiftly and noiselessly over their smooth expanse would eliminate a greater part of the nervousness, distraction and strain of modern metropolitan life.

Autos did spare twentieth-century man from being buried under the piles of horse manure predicted by an earlier century's doomsday prophets. Yet cars are nightmares for the mechanically disinclined. Horses tied up a large part of our work force a century ago. Cars do the same today. We construct and maintain a million miles of hard-surface highway and engage a sixth of our work force building, servicing, and operating our motor vehicles.

Buggies versus cars, farmlands versus suburbs, country stores versus supermarkets. Are we really better off?

Probably so. However, we may have lost a precious quality of life in 1876. Nineteenth-century man worked hard, suffered diseases, stayed most of his life in one place, and died young, but history and literature indicate that he also enjoyed life. He knew how to converse; he was neighborly; he experienced nature, the change of seasons, and the silence of the woods. We have better health and education, but our affluence has come with a price tag: smoke-laden skies, oily water, congested streets, poor folks in a land of plenty. Nevertheless, we have conquered the material elements, and that gives us hope.

2 · Consumerism Versus Environmentalism

Most Americans are overconsumers. They consume in ever larger amounts food, clothes, fuel, and hundreds of items that are used for a brief period and then discarded. Assembly lines and belching smokestacks are as much a part of our heritage as apple pie and the Fourth of July. We are accustomed to a plentiful supply of low-priced material goods, produced by an economy of ever increasing production and consumption, whose management we entrust to special business and industrial interests. In recent years, however, some citizens have banded together to protect the special interests of consumers. They are advocates of *consumerism*.

Consumerism does not connote the same thing to producers that it does to consumers. To the producer it connotes not only the making of wanted goods for profit but motivating consumers to purchase new products and winning customers away from producers of competing goods. It means advertising, sales appeal, and the highest quality product at the most profitable selling price. The producer makes clothes that will not fall apart at the first washing because he wants the buyer to return at some time in the future, but he makes clothes that last only so long—or else the customer would never return. The producer desires a good reputation, consumer loyalty, and product name recall. He also wants to convert newly made luxuries into consumer necessities as quickly as possible.

To the average citizen consumerism connotes a reasonable

choice of products, honest advertising, clear labeling, and product safety. Consumers are a varied lot. They are not specialists like auto makers or food processors, who readily band together to block emission control standards or to raise prices. They generally wield less political power than their special-interest adversaries and seldom have at their disposal the array of facts and statistics that would enable them to mobilize on particular issues. Because of this they tend to depend on governmental intervention, such as antitrust legislation, substantiation of advertising claims, regulation of unit price, and safety testing procedures, to protect their interests.

The differences in viewpoints and expectations on the part of consumers and producers are not too great. Both groups belong to the cult of commercialism and share a common language: high quality, product acceptability, longer life, bargains, and panic buying (in which the customer's resistance is broken and he is swept into action).

Consumerism from the broader sociological standpoint, in contrast, includes the producer's responsibility for the safety and welfare of the worker who makes the product. It includes the consumer's responsibility for the product when it is eventually discarded into the environment. It means that the producer might have to sacrifice profit and that the consumer might have to sacrifice convenience so that no harm comes to worker or neighbor.

Two Consumer Types

Consumerism may not be good environmentalism. High-quality detergent motor oil may be admired by drivers, but it defies traditional recycling techniques, and each year a billion gallons of such oil is dumped into ditches and waterways as waste, to the consternation of the environmentally conscious. Environmentalists as a rule want the internalized costs in commodities to reflect the true cost of energy expended, disposal, and resource depletion.

There are two types of consumerists—the consumption-production, or C-P type, and the consumption-environment, or C-E type. Often the two types demand the same things—longer-lasting products, lower prices, conservation of natural resources—but generally one type or the other predominates. C-P consumers tend to be frugal, mindful of the quality of the goods purchased, and quite concerned about personal health and welfare. They are usually wage earners or members of wage-earning families. Budget-conscious, they take their time in making purchases and aren't afraid to return defective or unsatisfactory items, no matter how small the price. C-E consumers are often more affluent than the C-P type. Well read, they are inclined to worry less about bare necessities and more about broader social issues. They devote time to community causes that do not immediately result in personal benefit and are interested in preserving natural surroundings. Although both types of consumers take public stands on issues, the C-E type finds it harder to muster support, since the issues they support are sometimes remote from the daily concerns of the average citizen.

Just as not all people concerned about consumer products are concerned about the environment, so not all environmentalists are concerned about consumer issues. The notion of limiting economic growth is the point at which C-P and C-E consumers part company. Both types are interested in consumer products, but they differ in their responses to growing shortages. The C-P folks want reasonably priced gasoline and meat and claim that limiting production drives up prices. C-E people believe that the full social costs of consumer products can only be paid by restricting production and consumption of limited natural resources.

C-P consumers have blind faith in America's technological ability to correct our self-made mess. Such a consumer may back global solutions to the exploding population problem but at the same time feel that pressure to make him participate in a car pool or use mass transit instead of driving his own car to work is a violation of his constitutional rights. The C-E consumer has less faith in technology. He may dislike restrictions on his driving practices but is willing to exchange such restrictions for cleaner air and less noisy streets. He drives a small car and eats less meat than the C-P consumer and does not believe that flashy gadgets are necessary for a better lifestyle.

The Arab oil embargo in the fall of 1973 created a mini-crisis for the C-P person. He was harassed by long lines of cars at gasoline stations, lack of accurate information from the oil companies, rapidly rising prices, and threats of layoffs. Both types of consumers now realize that there is something wrong when producers embellish their advertising with appeals to national pride, economic health, and job security. They realize that the battle is not consumerism versus environmentalism at all but commercialism versus environmentalism.

Even without the growing shortages of raw materials, producers face new challenges from C-P consumers in the form of increased auto recalls and improved safety devices. Producers admit that shorter-lived products require more raw materials, but such products, they say, return larger profits and engage a larger work force and thus help the economy. The C-P reply is that higher-quality products are safer, need less maintenance, wear longer, and in the long run use only half the materials of shorter-lived products. They may cost more, but they are better investments.

The Overconsumers

The U.S. citizen's consumption of materials is unparalleled in the history of the world—more than twice that of Western Europeans and four times that of Eastern Europeans. Now Canada, Australia, and a strip of nations stretching from Ireland through Europe and Russia to Japan are imitating our consumption habits, so that one-third of the world's people now consume two-thirds of the world's minerals and energy. Only a handful of Third World countries, such as oil-rich Saudi Arabia and Kuwait, can ever hope to belong to this exclusive consumption club.

The amount of materials consumed in the United States each year staggers the imagination—twenty-eight tons per person from mines, fields, forests, and oceans. The Environmental Protection Agency tells us that about 10 per cent of this represents agriculture, forestry, fishing, and animal husbandry, about 35 per cent fuels, and 55 per cent metals and other minerals and construction materials. Consumption is not uniformly distributed by any means. Our own "Third World"—the Appalachian region and the urban poor—do not consume nearly as large an amount of materials as the people in Scarsdale, New York, for instance, or those who live on the Miami Gold Coast.

For decades Americans have been skimming the cream off the

earth's resources. The U.S. has pumped over 100 billion barrels of oil out of the earth. If the rest of the world consumed at the same rate as Americans, they would exhaust and devastate the world's known material resources in a short time. The forests would disappear in a single decade if every human being were allotted one magazine and one roll of toilet paper each week.

America feeds its exploding pet population—increasing at the rate of 4 per cent per year versus 1 per cent for human beings—more protein each year than is consumed by an equal number of people in Bangladesh. If the rest of the world were to achieve the steel inventory of the ten richest nations, some 200 billion tons of iron—far in excess of the known world reserves—would have to be extracted from the earth's mines.*

Traditional U.S. optimism leads many of us to expect food and materials production to keep pace with human population growth. But if we think technology can eliminate poverty, we are in for an unpleasant surprise. The rich certainly live and eat better than their forebears, even if the poor of the world do not.

What are the expensive foods that the affluent countries are consuming in increasing amounts? They can be divided into two classes: empty foods and animal protein foods. Empty foods include the Crunchies, Burpies, and Junkies that American childern know as between-meal snacks. Overpackaged, overcolored, artificially flavored, and empty of genuine nutrition, they are a highly profitable portion of the $150 billion American food industry. They are environmentally costly, due to transportation and packaging.

Proteins are essential for human nutrition. Man can get these in meat, milk, and certain plants like soybeans. Americans get much protein from meat, and our country is the world's largest beef producer and importer. Our appetite has grown with affluence, going from about 50 pounds of beef annually per person in the 1940s to 113 pounds in 1971. A pound of beef requires grassland, cereals, and hay for fattening the steer and nourishing

---

* John L. Mero, in *The Mineral Resources of the Sea* (1963), estimates that the Pacific Ocean nodules alone contain copper equal to 600,000 years of annual U.S. consumption, nickel equal to 150,000 and titanium equal to 2 million years, but the high cost of extraction renders these resources inaccessible in the opinion of the present writer.

the brood cow. The cow takes a few years to mature and produces but a single calf per year thereafter, each of which requires another two years of tender loving care before it is ready for the market. Altogether it takes about a ton of cereal to make the food and meat each American consumes every year, in contrast to the 400 pounds of cereal consumed by the average inhabitant of the Third World.

Japan, while consuming 5.1 pounds of beef per person in 1971 —twice the amount of a decade before—still consumes only one twenty-second of the beef that Americans do—although the Japanese imitate the United States in such matters as baseball and Western folk music. The average Japanese eats 9.5 pounds of poultry annually (about one-fifth the U.S. rate) and 11.2 pounds of pork, compared to our 73 pounds. Nevertheless, the Japanese are immense consumers of protein compared with other Asians. The principal source of this protein is fish, which is eaten at an annual rate of 72.5 pounds per person, compared to 14.2 pounds in the United States. Between 1950 and 1968 fish consumption in Japan increased by about 5 per cent per year, but Japanese and Russian floating factories have stripped the nearby oceans to such an extent that global fish catches have leveled off and are beginning to dip.

Increasing competition for the limited world reserves of protein has contributed to major food shortages created by a succession of poor harvests in Russia and in Third World countries in the early 1970s. When Russian grain succumbed to severe winter weather due to a lack of snow cover, subsequent Soviet grain purchases wiped out the U.S. grain stockpile and drove food prices higher. Severe drought in parts of India and in the Saharan and sub-Saharan zone of Africa from Mauritania to Ethiopia has had an adverse effect on the world's food supply, as have conditions in Bangladesh, Indonesia, Korea, and the Philippines. Countries that normally export food, such as Nepal and Brazil, now import part of their food supply.

Technological miracles of the last few decades did buy time, but they are no permanent solution. The most publicized of these miracles is the so-called Green Revolution, which had its origin in Mexico in the 1940s when Dr. Norman Borlaug developed high-yielding dwarf wheat, which increased yield 300 per cent per acre in twenty years. The International Rice Research Institute in the Philippines developed a high-yield short-straw rice, but Asians don't like its taste and complain that it turns rock-hard when cold. It's good for famine relief and little more. The Green Revolution is further imperiled by the energy crisis, since high-yielding grains are dependent on nitrogen fertilizer, which requires methane gas for feedstock, and methane gas is in short supply due to the energy crisis.

A glaring example of overconsumption in the United States is the manner in which the U.S. use of the automobile is gobbling up the world's supply of mineral and fuel resources. In the United States the proportion of cars to people is higher than in any other country—one car for every 1.8 persons (Table II). Ten per cent of our material resources are expended in the construction and upkeep of these cars, and 15 per cent of the world's petroleum production is required to keep them running.

TABLE II
NUMBER OF MOTOR VEHICLES IN THE WORLD, 1971 AND 1972

	1971	1972	Population per Vehicle (1971)
United States[a]	112,999,125	118,618,200	1.8
Japan	19,857,877	22,408,500	5.2
West Germany	16,758,529	17,649,800	3.7
France	15,150,500	15,975,400	3.2
United Kingdom	14,276,682	14,909,600	3.8
Italy	12,291,700	13,516,100	4.4
Canada	8,306,418	9,052,400	2.6
Russia	5,400,000		45
Australia	5,049,800		2.5
Brazil	4,015,000		23
Spain	3,584,715		9.0
Netherlands	3,156,000		4.2
Argentina	2,539,000	2,670,000	10
Sweden	2,513,089	2,617,900	3.2
Belgium	2,399,584		4.0
World	262,174,539		14

[a] The United States has 43 per cent of the world's motor vehicles and 45 per cent of the world's automobiles.

SOURCE: "World Motor Vehicle Data," Motor Vehicle Manufacturers Association of the United States, Inc., 1973. Complete 1972 figures were not available.

Such consumption is not without its penalties. Gasoline-powered motor vehicles cause about one-half of our man-made air pollution. In fact, 64 per cent of carbon monoxide, 46 per cent of hydrocarbons, 37 per cent of nitrogen oxides, and 90 per cent of airborne lead come from gasoline combustion. Efforts to reduce this pollution have been nullified by rapid increases in fuel consumption. Other societal costs include $6.4 billion in cleaning bills and deterioration of homes and agricultural products. The greatest cost, however, is the total of 55,000 highway fatalities each year, which defies economic computation.

Motor mania is not restricted to the United States. Japan, West Germany, Italy, France, Great Britain, and Canada are close be-

TABLE III
ENERGY CONSUMPTION, DEVELOPED COUNTRIES, 1971

	Total Energy (*In thousands of short tons of coal equivalent*)	Pounds of Coal Equivalent per Capita
United States	2,565,758	24,789
Canada	221,981	20,560
Czechoslovakia	105,733	14,583
German Democratic Republic	118,486	13,907
Belgium	67,869	13,483
Sweden	54,399	13,424
United Kingdom	338,053	12,141
Australia	75,188	11,814
Denmark	29,156	11,744
Federal Republic of Germany	352,802	11,515
Norway	22,366	11,440
Netherlands	73,733	11,175
USSR	1,225,967	9,998
Poland	157,893	9,643
Finland	22,377	9,555
Iceland	981	9,504
Bulgaria	37,930	8,882
France	222,091	8,660
Switzerland	24,945	7,881
Hungary	37,599	7,255
Ireland	10,758	7,242
Japan	376,876	7,202
TOTAL	7,812,800	4,200

SOURCE: *Statistical Abstract of the United States,* Washington, D.C.: U.S. Government Printing Office, 1973, p. 815.

hind (see Table II). The motoring appetites of these countries are growing and will heavily drain the cheaper reserves of iron, copper, lead, and oil before reaching our levels of use.

In addition to foods and motor vehicles, there are other areas of excessive U.S. consumption. Every time we turn on a water tap, a television set, or an air conditioner, energy must be used to generate electricity to operate pumps and household appliances. Americans use twice as much energy as the British, three times as much as the French, and eight times as much as the average non-American (Table III). Between 1950 and 1970 the U.S. population grew from 152 to 204 million. During the same period electricity sales jumped from 381 to 1,391 billion kilowatt hours. Gross and net energy consumption is shown in Figures 1 and 2.

Only a few Third World nations (Nigeria, Algeria, Egypt, Libya, the Middle East, Indonesia, and five Latin American countries) have sufficient low-priced energy sources to satisfy their own needs, but their energy consumption is growing rapidly. Nigeria, with 60 million people and access to large domestic petroleum reserves, is increasing its use of petroleum at the rate of about 9 per cent annually. Such Third World nations are the fortunate minority. About 85 per cent of the Third World nations are energy-poor. They have neither fuel reserves nor the financial resources needed to drill oil under the ocean or to develop solar or tidal energy sources. These Third World countries face financial disaster because of rising prices of fuel, grain, manufactured goods, and fertilizer. India must trim its agricultural development program because of high prices and lack of fuel for tractors and irrigation pumps and the tripling of the price of urea, resulting in the loss of millions of tons of grain.

World stability demands that the industrial and energy "haves" share with the "have-nots." Without an international cooperative program the prospects for Third World development are nil. Some countries, like Brazil and Zaïre, will not be as badly hit because of increasing prices on their export commodities. Others, like India, Bangladesh, the Philippines, Sri Lanka, several countries in Central America and the Caribbean, and the countries of West Africa, are in real trouble.

Some energy-poor Third World nations try to keep their economies afloat by industrial development. Korea, Taiwan, and

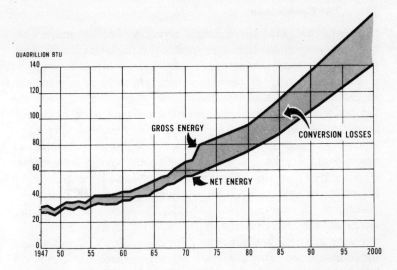

FIGURE 1. U.S. gross and net energy consumption, 1947–2000, in quadrillions of British thermal units, showing how losses due to converting fuel to usable energy such as electricity are expected to increase during the next three decades.

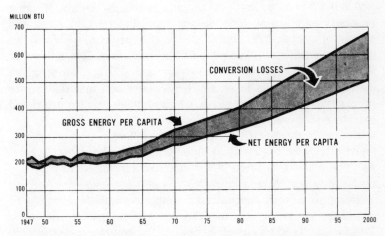

FIGURE 2. U.S. gross and net energy inputs per capita, 1947–2000, in millions of British thermal units.

Source for Figures 1 and 2: Walter G. Dupree, Jr., and James A. West, *United States Energy Through the Year 2000,* Department of the Interior, U.S. Government Printing Office, Washington, D.C., December 1972.

Singapore are examples. Brazil is trying to do the same. Since 1968 its gross national product growth (10 per cent) has been remarkable for a developing country, but growth has been achieved through the sacrifice of human freedom. The military government, which encourages this industrialization policy, has allowed the top 1 per cent of the population to increase its share of the national income from 12 to 17 per cent, while 50 million Brazilians at the other end of the income scale (half the total population) have found their share of the national income since 1968 cut from 17.6 to 13.7 per cent. During the same period the purchasing power of the average worker's salary has dropped 38.3 per cent.

Affluent nations are preoccupied with their own problems, which are largely consumption-induced: urban sprawl, air pollution, the possibility of nuclear mishaps, trade deficiencies, and traffic congestion. The more such nations consume, the more self-interested they grow and the more difficult it becomes for them to concern themselves with the problems of poor nations.

Contrasumption

American consumers are rapidly becoming aware of what shortages of material things mean. The blind faith in a plentiful supply of commodities and a lovely environment—provided we drive a few miles to reach it—has been shaken. Our notions of individual freedom, our right to purchase and use whatever we desire, and our belief in unlimited resources and material growth have all been called into question. Those whose voices are raised against our production-consumption ethic are gaining courage. They are beginning to feel they really belong to the American scene. They are beginning to write and speak about *contrasumption,* the philosophy of curtailing material consumption by directly challenging our material growth ethic.

The arguments of the contrasumptionist become clearer when we compare two kinds of American consumers—people and cars. The average U.S. citizen consumes one ton of organic matter each year in the form of renewable energy sources, such as fruit and grain. But automobiles compete with people for the total organic pool of this earth, and the average car consumes nonrenewable hydrocarbons to the tune of three tons a year, mostly in the form of gasoline. Cars are therefore three times the consumers of organic materials that people are. Up to 1974 America's motor vehicle population was increasing at the rate of 3 per cent per year, or over three times the rate of increase in the U.S. population.

A great number of Americans deplore the rapid rise in human population in Third World countries. We often imply, by using the label "overpopulation," that the procreative rates of individual poor families and nations are the determining factor in causing the global social crisis and the principal barrier to successful change. *They* have to change. There is little *we* outsiders can do, so let's continue in good conscience to enjoy our relative comfort.

The "theys" of the world are always some distance from us either geographically or philosophically. "They" include Latin American lands like Venezuela, which have succeeded in reducing death rates but have retained high birth rates, and so have population increases of 3.5 per cent per year. But if we take the 2 million yearly increase in population in the United States (neglecting other energy consumers such as pets) and add to that the increase in the number of motor vehicles, which consume three times as much organic materials as humans (equivalent to a 9-million increase in U.S. population), we have a total *equivalent* population increase in the United States of about 5.5 per cent per year. Comparing this to the actual population increase of 3.5 per cent in Venezuela, it becomes apparent that Americans are consuming organic resources at a far greater rate per capita than Venezuelans. Yet official U.S. policy is more concerned about Venezuela's need for contraception than our own need for contrasumption—and this at a time when we are rapidly depleting Venezuela's oil reserves to satisfy the thirst of our cars. This does not escape the attention of Third World radicals, who are quick to point out the hypocrisy of our population and consumption policies.

These policies are common to most developed countries. The motor vehicle population of Western Europe is presently increasing at the rate of 8 per cent per year. These European nations have fewer cars per capita than we do, but they are closing the gap, and absolute increases amount to 5 million cars per year, or 2 million more than in the United States. Human population growth is almost stabilized in most of these countries at 0.5 per cent per year. The motor vehicle increase, however, is equivalent to a human population increase of 15 million, or 6 per cent per year, almost twice the actual population growth rate of the most explosive Third World countries.

Herein lies the hollowness of our message. We preach birth control to the Third World and offer multicolored condoms and incentives for visits to sterilization clinics, but we are deadly silent about auto birth control at home. The United States has reached birth levels below replacement of 2.1 children per family but still aspires to two and three cars per family.

If contrasumption is to become a viable and accepted philosophy, it must include a global concern for the environment and a demand that we first put our own house in order. It must press for mass transit and bike networks, for countercommercials and for taxes on inefficient cars. If contrasumption were to become the order of the day, there might be an excise tax on the number of "mouths" per family—that is, two parents, one child, and *two* cars would be taxed the same amount as a family of two parents, four children, and *one* car. A more equitable excise tax would take into consideration the gasoline appetite of the car—that is, a 7-mile-per-gallon hog would be taxed at three times the rate of a 21-mile-per-gallon small car.

The youth culture is rapidly becoming contrasumptive. Blue denims and work shoes, family bike rides on Sundays and holidays, organic and homegrown foods, communes and rural life are all part of the contrasumptive culture. Work clothes last longer and are more comfortable (this has cost the clothing industry about $6 billion over the last few years); bikes cost less than cars; a vegetable patch saves on the grocery bill. Communal living, which involves shared use of washing machines, heaters, and other utilities and appliances, is good energy economics. The number of dwelling units and the high cost per unit have more to do with residential energy expenditure than does the number of persons per family. Seventy million household units of three members apiece take more energy than 50 million five-member units (but the latter has 40 million more members). A move to communal living would mean major savings, but actual present trends in this country, unfortunately, are toward smaller households.

In some communal residences the need for an electric dishwasher is eliminated by extending communal participation into the clean-up period. Americans, however, might be slow to accept the savings represented by the Japanese-style communal bath. Michael Corr and Dan MacLeod, writing in *Environment* in 1972, estimate that such a bath takes about 17,300 British thermal units per person, or the same as a small bathtub. An electric water heater and a large bathtub require 81,000 Btu's per person. Still more efficient than the communal bath would be a 5-minute light shower (5,850 Btu's) or a gas-fired sauna and cold shower (4,400 Btu's).

Organic- and homegrown-food faddists are against buying food at supermarkets. They close their ears to proponents of commercial agriculture, who use chemical fertilizers and pesticides, preferring biological control of insects, fertilizer of mulched garbage and leaves, and sun-ripened—though imperfect—fruits and vegetables. They are opposed to nitrate and phosphate fertilizers that enter waterways and add fluoride contaminants to the soil. They object to additives in foods and are suspicious of the lucrative food processing industry. Their swelling numbers include vegetarians, advocates of uncooked foods, and habitués of organic food restaurants.

The most fertile ground for recruitment into the contrasumptive ranks is the body of conservation-minded groups and individuals who are engaged in desperate attempts to stop the land developers and preserve some of our few remaining historic, scenic, and wilderness sites. Public-interest groups are another fertile field for recruitment, especially those concerned with corporate responsibility, occupational health, countercommercial work, and community organization and legal services. It is difficult to predict which public-interest groups will become contrasumptive. Much depends on leadership, group dynamics, and sources of funding. The major determinant is funding since few foundations that owe their existence to corporations profiting from the American growth ethic will support contrasumptive causes. The flux occurring at this moment among public-interest groups makes it impossible to rate them according to contrasumptive philosophy. However, some general observations are in order:

1. Groups that will not likely become contrasumptive in official policy include labor unions in which there is severe pressure from the rank and file membership for expanding production, consumer

research and informational groups who have adopted the C-P consumerist philosophy, and traditional conservation groups with corporate funding sources.

2. Groups that will most likely become contrasumptive include environmental lobby, advocacy, and coalition groups and consumer groups who have adopted the C-E philosophy.

3. Groups that may go either way include many religious lobby groups, groups concerned with corporate responsibility, and professional lobby groups.

Major factors for producing pro-contrasumption changes in philosophy are imaginative leadership, publicity-potential issues, security in finances and funding (by small liberal foundations or private donations), and increased communication with like-minded international organizations.

Public-interest groups hold the swing votes in the contrasumption battle. Since they seek to influence public opinion, they gravitate to popular issues, but most contrasumptive issues do not fit that category. Contrasumers soon find that establishment money controls the primary sources of public information—broadcasting and the press—thanks to advertising money.*

Governmental agencies, such as the Federal Trade Commission, the new Consumer Product Safety Commission, the Food and Drug Administration, and the Environmental Protection Agency, include on their staffs persons who are gradually becoming committed to contrasumptive philosophy. This is producing good results. The Federal Trade Commission is showing an aggressive attitude toward suspected oil, drug, appliance, and auto advertising claims, the Food and Drug Administration is strengthening its standards for new drugs, and the Environmental Protection Agency is perfecting its program for monitoring industrial pollution. While these actions are not primarily contrasumptive, they do curtail consumption to some extent.

Some federal agencies, such as the U.S. Bureau of Mines and the Departments of Agriculture and Commerce encourage the material growth ethic, while attempting to regulate producers.

---

* A study of the media coverage of the late President Allende of Chile showed that many leading U.S. newspapers constantly referred to his lack of popular support. Only the *Christian Science Monitor,* which carries no advertising, gave a more balanced report.

This contradiction in function is even more blatant in the case of the Atomic Energy Commission and the Federal Power Commission, where regulation does little to interfere with growth of energy consumption. This country desperately needs a national conservation policy and an effective regulatory commission to see that such a policy is carried out.

The Deadly Game of the Ik

Anthropologist Colin Turnbull, in his provocative book *The Mountain People,* tells a terrifying story of a small African tribe called the Ik, who live in the mountains of Uganda. This formerly nomadic tribe has been penned in by national and regional boundaries and the fences of game reserves. Their environment was severely disrupted in a short period of time. By the time Turnbull visited their settlements, the entire social structure of the tribe had broken down. Children were driven from their homes at the age of three to scavenge for food. Adults paid little attention to each other, performed sex without passion, laughed at each other's sufferings, and raided and stripped the homes of the recently deceased. The Ik went for long periods without talking; they merely sat, watched, and waited for someone to bring them food.

The god of the Ik had died decades before. There was no ritual for marriage or burials, and the people had forgotten all social customs and traditions. Family and tribal bonds, usually strong in Africa, were lacking, and the population was simply dying out. When the government trucked in food supplies to the point where the road to Ik territory petered out, only able-bodied youths had strength to walk to the delivery points. Instead of packing food back to their communities, the young men gorged themselves, became sick, vomited, and then gorged themselves again. The needs of their relatives and friends meant nothing to them.

This story is not apocryphal. It is being enacted today in Africa, and elements of the same tragedy are to be found in the streets of New York and Calcutta. We all have some Ik in us. A traumatic break in our environment could drive many of us to become still more isolated and individualized in our actions. The fragile social structure that has taken man thousands of years to construct could break down under the weight of environmental stresses.

Commercialism, especially as practiced by the Western world, has put strains on our social fabric. We have become fenced in and congested; our water and air is being polluted; our sources of readily available energy and protein are becoming scarce; nations are scrambling for raw materials with little regard for the needs of other nations. The developed countries gorge themselves and vomit their wastes from smokestacks and sewer pipes, while the young developing nations are turned loose and advised to scavenge for themselves, although only the more energetic and developed among them are able to reach the more distant sources of materials.

The practice of sharing with other nations is frowned on by nations with shortages and environmental crises of their own. U.S. foreign aid is openly questioned with each new appropriations bill, and the percentage of the total national income allocated to foreign aid shrinks almost yearly. The race for extra oil reserves, for fishing grounds, for timber and minerals is deadly serious. In June 1973 the Chicago Grain Market suspended sales of soybeans, because there were no more futures to sell. At about the same time Icelanders were shooting at British fishing boats that came too close to their shores.

Rising food prices, the scramble for living space, cutthroat competition, embargos, panic hoarding, and inflation are becoming commonplace. The world goes from one crisis to another at great cost to society and the environment. In the meantime, ritual and symbols are being forgotten. God died decades ago for many, and respect for man is fading fast. Before we, like the Ik, follow this road to extinction, it behooves us to examine more closely our present crises in order to arrive at a rational course of action.

3 · Comprehending the Crisis

In April 1973 American Petrofina reversed the flow of petroleum in its pipeline between Midland, Texas, and Harbor Island in Corpus Christi. For some forty years the pipeline had carried crude oil out of the Permian Base in Texas to the Gulf of Mexico. The line now brings 10,000 barrels of imported crude oil daily *into* the heart of Texas.

The ominous reversal of the flow of a pipeline in the richest oil country in the world should give us pause. Is this part of a major crisis or is it a passing shortage? Back in 1970 the East Ohio Gas Company could not supply fuel to its customers in Cleveland, and about 35,000 workers were temporarily laid off. Fuel shortages have cropped up time and again in our history, but signs of a more serious and lasting shortage began to appear during the summer of 1972.

By the following winter severe weather conditions gripped parts of the nation, killing 50,000 head of cattle in the Texas Panhandle, shutting down schools throughout the Great Plains, and closing sugar mills as far south as Louisiana. Economic life was disrupted in the very heart of gas- and oil-producing country. In the upper Midwest the lack of natural gas, needed for the heaters used to dry the wet soybean and corn crop, added to agricultural losses and rising consumer prices. Luckily, in that section of the country the unseasonably mild winter forestalled a severe crisis.

In the spring of 1973 the breathing period that usually occurs between production of winter fuel oil and the production of summer gasoline was short-lived. The worst flood in U.S. history hit the Mississippi Valley, disrupting the barge traffic that carries crude and refined oil on inland waterways. The shortage was beginning to be felt.

As spring gave way to summer, the television commercials of the big oil companies changed their tune. Johnny Cash, Amoco folk singer, plugged conservation of gasoline, rather than consumption. Other companies followed suit. Predictions were made that the coming winter would witness still more severe shortages. Homeowners were urged to insulate their houses. In the fall of 1973 the Yom Kippur War was followed by the Arab oil embargo, aimed especially at friends of Israel. The consuming nations of the world were shocked. The spigot had been turned off by the Arab nations at the very time when importers had hoped to increase oil flow by several million barrels a day.

Surprisingly, during a time of peace and relative prosperity—and low Administration credibility—the people went along when the President pleaded for voluntary fuel conservation. The winter was mild, and the fuel oil lasted, but there was irritation over increased fuel costs and lengthening lines of cars at filling stations. A quarter of a million workers were temporarily or permanently laid off. Makers and dealers of big cars were especially hard hit, as overnight the gas guzzler fell from favor.

Within a matter of months during the fall and winter, things had changed radically for the consumer. Shortages had become a permanent part of American life. Gas wars, low-priced fuel, and big cars suddenly seemed like ancient history. Within the space of two years the United States had changed from a net exporter of fuel to an importer. Filling station closings were ten times the normal rate—10,000 out of 220,000 in 1973 alone.

When consumers recovered from the initial shock, they began the search for someone to blame. The major oil companies blamed governmental policies and environmental restrictions. The liberal wing of the consumer movement pointed the finger at Big Oil. There was plenty of fuel on tankers off the coast, they said, or in storage tanks or at the service stations. They used words like "conspiracy" and "price-gouging." Private and governmental investigators tried desperately to determine the true facts but found many of them lacking. The few facts available seemed to bear out consumers' accusations. The 1973 profits of the big oil companies were spectacular compared to those of the year before:

	1973 Profits	Increase
Exxon	$2,400,000,000	59.5%
Phillips	230,000,000	55.3
Standard of California	844,000,000	54.2
Mobil	843,000,000	46.8
Texaco	1,300,000,000	45.1

No single cause—environmental restrictions, lack of refinery capacity, continued U.S. export of petroleum products, failure to furnish tax incentives for exploration—could be called the dominant factor. The complexity of the problem angered citizens all the more. Big Oil held the data sheets and knew what was happening. The shortage—if not caused by Big Oil—was at least welcomed as a chance to break the independent oil companies and raise prices, to increase tax breaks and profits, to lower environmental restrictions, and to justify exporting refinery capacity to the Middle East, Europe, and the Caribbean.

Amid all the soul-searching, some blame had to be shouldered by consumers. A closer look at the figures would have alerted them well in advance. Gasoline consumption had jumped 7 per cent in 1972 to 102.6 billion gallons a year and then jumped another 5 per cent in 1973. Americans consumed 68.8 quadrillion British thermal units of energy in 1970, 72.1 in 1972, and an estimated 75.6 in 1973 (Table IV). The Department of the Interior predicted that, unless an energy conservation policy was established, U.S. consumption during the 1980s would be double that of the 1970s and three times that by the first decade of the new century (see Figures 1 and 2 in Chapter 2). Americans change residences about once in five years, commute great distances to work, and vacation in far-off places. In fact, 25 per cent of U.S. energy is used in transporting Americans and their goods from one place to another. More and more of our citizens migrate to moderate climates like Florida, Arizona, Texas, and California. We are becoming a hothouse people, preoccupied with keeping our homes warm in winter and cool in summer. We use another 25 per cent of our energy in heating and cooling residential and commercial

TABLE IV

U.S. Gross Consumption of Energy Resources, by Sources and Consuming Sectors, 1972–73

Consuming Sectors	Anthracite	Bituminous Coal and Lignite	Natural Gas	Petroleum	Hydropower	Nuclear Power	Total Gross Energy Inputs	Utility Electricity Distributed	Total Net Energy Inputs	Percentage Change from 1972
			In trillions of British thermal units							
Household and commercial										
1972	75	312	7,642	6,667	—	—	14,696	3,478	18,174	
1973[a]	75	295	8,001	7,024	—	—	15,395	3,727	19,122	+5.2
Industrial										
1972	35	4,232	10,591	5,668	35	—	20,561	2,493	23,054	
1973[a]	29	4,425	10,825	6,043	35	—	21,357	2,671	24,028	+4.2
Transportation										
1972	—	4	790	17,264	—	—	18,058	17	18,075	
1973[a]	—	5	814	17,927	—	—	18,746	18	18,764	+3.8
Electricity generation, utilities										
1972	40	7,797	4,102	3,134	2,911	576	18,560	5,988		
1973[a]	36	8,655	3,918	3,435	2,906	853	19,803	6,416		+6.7
Miscellaneous and unaccounted for										
1972	—	—	—	233	—	—	233		233	
1973[a]	—	—	—	260	—	—	260		260	
Total energy inputs										
1972	150	12,345	23,125	32,966	2,946	576	72,108		59,536	
1973[a]	140	13,380	23,558	34,689	2,941	853	75,561		62,174	+4.8

[a] Preliminary.

SOURCE: Division of Fossil Fuels-Mineral Supply, Bureau of Mines, U.S. Department of Interior, March 1974.

establishments. Industry accounts for another 32.2 per cent of the total energy consumed in the United States.

Whether the period of the 1973 Arab oil embargo will eventually be described as an energy crisis, a "critical period" (Secretary of Defense James Schlesinger's term), a temporary shortage, or a permanent condition, only history will tell. However, it was certainly a time when the consumer had to rethink a lot of things formerly taken for granted. He had to consider several well-defined consequences of the overdependence of industrialized countries on Middle East oil:

1. Increased fuel costs, resulting in enormous trade deficits, capable of wrecking the healthiest economies; no guarantee of continued increases in crude petroleum production from many oil exporters.
2. A shift back to coal for power plants; increased research on coal gasification and liquification; spread of strip mining to Western states.
3. Mounting pressure to drill for gas offshore, especially off Atlantic Coast states; pressure to deregulate natural gas versus efforts to make natural gas information public.
4. Continued pressure by the Atomic Energy Commission and the nuclear equipment industries for nuclear power plant construction; efforts by environmentalists for a moratorium on such power plants.
5. Stepped-up research on alternative energy sources, especially solar energy.
6. Mounting pressure to relax sulfur oxide emission standards; postponement of auto emission regulations to a later date.
7. The joining of a battle to divert military and space money to alternative energy technologies.
8. A growing awareness of the need for fundamental changes in the size and types of cars, for ways to curtail energy consumption and restructure the oil industry, and for establishment of a national energy policy.

Although the public has been able to obtain very few facts from the oil companies as to production, transportation, refining, and storage of fuel, shortages do exist. A major nonrenewable resource is being rapidly depleted, and critical energy decisions must be made now. This in itself is enough to justify calling our present predicament an "energy crisis."

Petroleum—Liquid Gold

Petroleum is the cornerstone of the industrial world's energy pool, thanks to the decline of coal production in recent decades and to the popularity of the automobile. Petroleum liquid products by the early 1970s supplied almost half of America's total energy needs and even higher percentages of the energy needs of Japan and many Western European countries. This growing dependence on a single energy source places consumer nations at the mercy of the oil-producing nations, who, in times of shortages, embargoes, and cutbacks in production, can dictate almost any price. This is exactly what happened in the fall of 1973, when the price of Nigerian, Arabian, and Iranian oil on the world market increased five- and sixfold.

Much has been learned about petroleum since it was first commercially pumped from the ground in the 1850s; however, much of this information has not been made public, such as the extent of oil reserves throughout the world, the amount of oil in storage at this moment, or the actual over-all profits of the petroleum industry. We are told that in the United States oil producers get only thirty-five barrels of oil per foot of exploratory drilling in the 1970s compared to 270 barrels in the 1930s, but who knows what is in store for us under the oceans or in the Arctic.

The cost of exploration and drilling in remote and inaccessible regions has skyrocketed but, once a producing oil well is established, the extraction process is quite cheap, as is the piping or shipping of liquid fuels. Black petroleum then becomes liquid gold. It requires no mining and can be refined into a number of fuel fractions that serve as valuable feedstocks for the petrochemical industry.

Until late 1973 many consumers were unaware of the disparity in petroleum reserves, with the Middle East possessing 57 per cent of the world's known reserves, worth such a fabulous amount that a handful of Arab sheiks are literally capable of breaking the world bank (Figure 3). A belief on the part of the consumer countries that the producer nations would maintain low oil prices for years to come was coupled with the naive expectation that oil would flow to the consumers in ever increasing amounts by simply

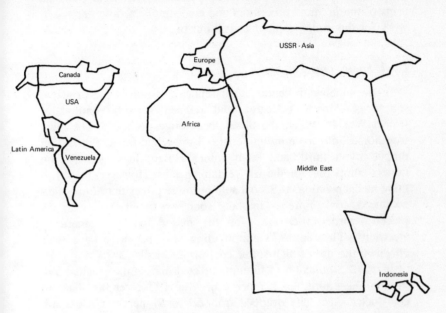

FIGURE 3. World geographic areas in proportion to oil reserves.

twisting a valve. Well into 1973 U.S. Government planners were speaking of increasing oil imports by 20 to 30 per cent a year in order to meet the country's growing energy demands.

The producers were the first to realize that they were in a seller's market when the major oil-exporting countries united on plans for curtailment of petroleum exports. Canada limited the export of oil to the United States in early 1973 and announced a policy of shipping its excess oil to its own energy-poor Maritime Provinces. Libya, Iran, Kuwait, and Saudi Arabia all developed plans for curbing oil production. Middle East production dipped from more than 20 million barrels a day in September 1973 to about 15 million by late fall.

Voluntary fuel-conservation measures and rationing are only temporary solutions to the problem, as consuming nations wait for new oil sources to be developed in the North Sea and Alaska.

In the meantime the United States is moving away from an oil-dependence posture as rapidly as possible—and right into the arms of the uranium processors and coal suppliers, who happen to work for the same people that drill our oil.

Coal, Shale, and Tar Sands

In the nineteenth century coal replaced wood as the principal source of America's energy and reigned supreme until after World War II. When the diesel locomotive pushed the belching iron horses into the museum, coal fell upon bad times. Production dipped by a third, and many mines closed, leaving 1 million miners competing for the few remaining jobs. However, thanks to mine mechanization, U.S. coal sold at lower prices than European-produced coal. Foreign markets soon expanded, and coal replaced hydroelectric power as the major fuel for generating electricity. Then in 1970 antipollution laws began to have their effect on the use of high-sulfur coal in power plants.

In 1972 production of bituminous coal and lignite reached 590 million short tons, up 6.8 per cent over 1971, yet the share of coal-fueled electricity dropped from 52 to 51 per cent. Coal still remains our only exported fuel, with 57 million tons exported in 1972; 338 million tons were used to fuel our electric power plants in the same year, and 157 million tons were mined for industrial uses in coke plants and steel and rolling mills.

The oil embargo of 1973 and subsequent talk of energy independence by 1980 was music to the ears of the coal producers. The United States possesses over 3 trillion tons of coal and expects to mine only one-half of 1 per cent of these reserves in the coming decade. (Anticipated quantities are projected through 1985 in Table V.) In 1973 pressure began to mount to convert some power plants from oil and gas back to coal, and electric power groups continued pressure for softening sulfur dioxide emission standards.

The emphasis on coal as a partial solution to America's energy problem is by no means altogether welcome. More and more of our coal production has been coming from strip mines, which prove more profitable to the producers than to the environment. Mechanical shovels, sometimes eleven stories in height, scoop immense

TABLE V
U.S. Coal Estimates, 1970–85

	1970	1975	1980	1985
	In millions of short tons			
Conventional uses	519	662	852	1095
For export	71	92	116	138
Synthetic fuel:				
Gas from coal	–	–	48	232
Liquids from coal	–	–	12	107[a]

[a] Equivalent to 680,000 barrels of oil per day.

SOURCE: *U.S. Energy Policy,* National Petroleum Council, Washington, D.C., 1973.

quantities of coal from near-surface locations with only a fraction of the workers required to maintain underground mines. Each week 2,000 acres (100,000 a year) are gouged out. Strip-mined states are becoming technologically induced wastelands. By 1980, 5,000 square miles, an area the size of the state of Connecticut, will have been stripped. Reclamation laws seem to go unenforced, and fines fail to cover the cost of returning the land to its original contours and productivity. Coal seams have had the effect of retaining moisture in many places (as in North Dakota), and with their removal the land becomes a desert incapable of growing crops.

The coal-mining center of this country is shifting from the Appalachian region to the Western states, which have large deposits of easily strippable coal. Part of this shift is because about 40 per cent of the coal east of the Mississippi has 3 per cent or more sulfur content, and methods for desulfuration are not well enough advanced to make high-sulfur, deep-mined coal competitive with less-polluting, surface-mined Western coal.

Hand in glove with plans to exploit the vast U.S. coal reserves go those for developing the shale oil formations in our Western states and the tar sands of Western Canada. Trillions of tons of liquid fossil fuel are locked up in such formations, but the technology for recovery of oil from shale rock is in its infancy. A barrier to making shale oil a major contributor to the energy picture is that some of our richest deposits are in water-short regions of Colorado, Utah, and Wyoming, and shale oil extraction requires about five barrels of water for each barrel of oil produced. Many of the environmental problems associated with stripping coal can be expected in shale oil regions, which are among the most scenic in our land.

Tar sands are found mostly in Western Canada. In northern Alberta the miles and miles of spongy bogs around the Athabaska River cover one of the richest oil reserves in the Western world—deep deposits of gummy black sand that may contain as much oil as Saudi Arabia. This is no solution to U.S. problems, for the Canadians have placed severe restrictions on exports of oil to the United States, and this applies to tar sand products as well as crude oil.

Natural Gas—Clean but Lean

The United States burns over half of the natural gas tapped and collected in the world each year. Little wonder that our own reserves of this fuel continue to diminish. Studies by the American Association of Petroleum Geologists estimate that twenty to forty years of "probable" U.S. gas reserves are left. These dire predictions are sad tidings for gas consumers who use this clean, cheap fuel for kitchen ranges and home heating. Unfortunately, so do power plants, which find natural gas a good substitute for high-sulfur coal and oil. That there is not enough gas for everyone was brought home emphatically in early 1973 when the people of San Antonio and Austin saw their electric power cut—right in the heart of the Texas gas and oil country.

While total energy consumption in the United States rose 4.9 per cent in 1972, natural gas consumption increased a modest 2.1 per cent compared to 3.6 per cent in 1971. In 1972 there were 22,607 billion cubic feet of gas produced, of which a third went to households and commercial outlets and about half to industrial outlets. This immense appetite for natural gas has led to problems, and many suppliers now refuse to take new customers. Gas companies blame low prices of natural gas for continued declines in exploration and drilling and say that raising gas rates will encourage the search for new reserves. Here the picture becomes a little confused.

In the summer of 1973 Senator Philip A. Hart of Michigan

called the natural gas shortage a "hoax warranting congressional action." His Antitrust and Monopoly Legislation Subcommittee staff subpoenaed producer records showing reserves of gas up to 1,000 per cent greater than the firms had been consistently reporting to the American Gas Association. The companies have lobbied for decontrol of interstate prices, so that investors could put more money into exploration. However, Monte Canfield, Jr., deputy director of the Ford Foundation's Energy Project, testifying before the Senate subcommittee, said new gas will come mainly from offshore fields, which are prized for oil but almost always contain gas. The exploratory drilling is not limited by investor money but by the availability of oil-drilling equipment.

Traditionally, gas discovery has been dependent upon oil-exploratory activity. The frantic search for oil in many parts of the world should lead to new natural gas finds. Petroleum drilling in the North Sea is expected to discover enough gas to supply much of the future needs of Northern Europe and Great Britain. The Netherlands has discovered enough natural gas since World War II to supply its total domestic market and is still able to export gas to other Common Market countries. Algeria is another known source of natural gas. There is serious talk of shipping liquid natural gas in tankers to the United States. However, to ship enough gas to meet our growing needs would require forty-five tankers priced at $100 million apiece, each carrying a combustible cargo so potent that, if triggered by an errant spark, it would destroy a large portion of any port in which the tanker docked. Liquid natural gas disasters are not unknown. In 1944 in Cleveland, Ohio, a liquid natural gas explosion leveled some thirty city blocks and took 180 lives.

The potential danger has not halted businessmen in their frantic search for foreign supplies of gas. One untapped reserve is in Siberia. The Russians, hungry for American grain and machinery, are aware of our appetite for natural gas. Apparently some Russian planners hope to tie our energy needs into their own development programs. It takes enormous outlays of capital to explore and deliver natural gas from remote Siberian fields. This temptation to use Western capital has been resisted by Soviet conservationists desiring to retain energy resources within the borders of the U.S.S.R. By mid-1974 it was becoming apparent that the Russians would renege on contracts to deliver 400 billion cubic feet of gas per year to West Germany, France, and Italy in the late 1970s. They are also having second thoughts about Japanese and American participation in development of Siberian natural gas reserves.

Natural gas is clean and easy to transport, and once people begin to use it there is great resistance to change to another fuel. Inevitable higher prices of natural gas are going to make synthetic gas from coal more competitive. Coal-gasification technology is not new but would demand large quantities of coal. El Paso Natural Gas Company and Texas Eastern Transmission Corporation intend to build coal-gasification plants in the Four Corners area of the Southwest. These plants will cost $400 million apiece and produce a quarter of a billion cubic feet of natural gas per day, or only 0.4 per cent of current needs. The coal needed to meet our gas requirements is an astounding 6 billion tons per year, ten times our current coal production.

Coal gasification and importing of liquid natural gas will come as no bargain. They require increased prices of existing supplies in order to be competitive, and there is pressure from vested interests to see that this occurs. Coal-gasification demands could turn the American landscape into a wasteland. In the production of synthetic gas, hydrogen from water and coal combine to form a gaseous product that contains less than half the energy of natural gas. Synthetic gas can either be burned directly for making electricity or processed further to yield pipeline-quality gas. Such processing is expensive. If the cost of adequate strip-mine reclamation is taken into consideration, synthetic gas will cost over $1.25 per 1,000 cubic feet (the average 1973 price was 25 cents).

Nuclear Power—Tarnished Panacea

The earth's supply of fossil fuels is fast dwindling, and arguments for the use of nuclear power sources are more appealing than ever to a nation whose continued wealth and well-being depend on energy. Riding the crest of the energy crisis, the Atomic Energy Commission (AEC), nuclear reactor manufacturers, and a growing segment of the electric utility industry have mounted a campaign to convince the public that the atom is essential to future energy needs.

The arguments at first glance seem compelling. In 1973 the Federal Power Commission predicted that by 1990 the electric power industry will require installations of nearly 1 million megawatts generating capacity to meet national needs, compared with installations producing over 400,000 megawatts today. The commission further predicted that by 1990 nuclear power plants will have to meet fully half of the nation's power needs compared with only 4 per cent today. At the beginning of 1974 there were only thirty-seven nuclear power plants with operational licenses. The Atomic Energy Commission estimates that there will be 900 in operation by the year 2000.

Nuclear power advocates make a strong case for using untapped uranium reserves. Reserves are estimated to range from 273,000 tons of uranium oxide ore at eight dollars a pound to about 1.6 million tons at fifteen dollars a pound—the cost increasing with the decrease in uranium concentration. Like fossil fuel reserves, uranium deposits are a limited, nonrenewable resource. The new fast breeder reactors now on the drawing boards, which will allow 60–70 per cent utilization of energy content of uranium, are more a dream than a reality. In the meanwhile, nonbreeder reactors currently in use utilize only 1–2 per cent of uranium's energy potential.

This fantastic inefficiency aside, environmentalists are gravely concerned about the by-products of all nuclear power production —radioactive wastes, which are potentially dangerous for thousands of years. To a population accustomed to disposing of its wastes down the nearest sinkhole, such long-term problems are incomprehensible. It seems easy enough to just drop the stuff in the ocean or bury it deep in the Permian saltbeds in the Kansas heartland, which the AEC assures us is perfectly safe.

But safety here depends less on eventual disposal and more on intermediate disposal sites and transportation to the final nuclear graveyard. This is what has environmentalists worried. Tank leakages in disposal depots, while not frequent, still happen. In June 1973, 115,000 gallons of radioactive waste leaked into the ground at the AEC's Hanford Reservation in Washington State.

Since 1958 there have been fourteen tank leaks, and 300,000 gallons of radioactive waste have been dumped into the nation's soil. The Hanford leak occurred on high ground seven miles from the Columbia River, where the water table is from 250 to 300 feet below ground. The AEC's repeated assurances that such spills are not dangerous must be taken with a grain of salt when so much is at stake.

The result of a nuclear waste disaster or reactor failure would not, as many believe, be an explosion like the one that leveled Hiroshima. The danger lies, rather, in atmospheric and environmental contamination of highly toxic radioactive materials, the most dangerous of which is perhaps plutonium-239, derived from uranium in the course of the reactor operation. One pound of this material has the potential to cause lung cancer in 9 billion human beings—two and one-half times the current world population; a "spill" could make an entire region hazardous for a thousand generations. A guarantee of 99.999 per cent containment is simply not enough. The remaining 0.001 per cent, if uniformly dispersed, could cause severe harm to all the world's people.

For these reasons optimistic statements about vast uranium reserves, efficient nuclear power plants, and safe burial grounds for radioactive wastes have failed to convince increasing numbers of concerned citizens. A vocal organization made up of some 115 citizen groups under the name "National Intervenors" points out that, contrary to public reassurances that the chance of a disastrous power plant accident is small, electric utilities have found it necessary to protect themselves from liability if one occurs. The Price-Anderson Act, as the National Intervenors have pointed out, acknowledges that catastrophic nuclear accidents can happen but exempts the utilities from full liability for public damages.

In addition to its pitch for public credibility, the nuclear power lobby is demanding federal assistance in the following forms: long-range uranium purchase contracts between producers and utilities; uranium selling prices that will include the cost of discovery, development, and production plus a reasonable rate of return on investments; continuing favorable tax treatment; and access to public lands, as necessary, for uranium exploration and development.

Citizens are now asking for some simple federal assurances and safeguards for their own—not industry's—health and protection. They resent the AEC's continued attempts to foist unsafe radiation standards on the public when there is no evidence whatsoever that there is any safe threshold for radiation exposure. Walter Jordon, a pronuclear member of the Atomic Safety and Licensing Board, admits:

> The important question still remains. Have we succeeded in reducing the risk to a tolerable level, that is, something less than one chance in 10,000 that a reactor will have a serious accident in a year? . . . The only way we will know what the odds really are is by continuing to accumulate experience in operating reactors. There is some risk but it is certainly worth it.*

The heart of the problem is: The risk is simply not worth it. John Gofman of Berkeley, a longtime critic of the nuclear power industry, asserts that when 500 nuclear reactors are in operation —about half the number expected by the year 2000—we could assume one major accident every twenty years. Such an accident, he claims, could wipe out a city the size of New York—hardly a "tolerable risk." Most citizen groups go along with Gofman when he says, "The issue is not technical but moral—the right of one generation of humans to take upon itself the arrogance of possibly compromising the earth as a habitable place for this and essentially all future generations."

Concern about nuclear power safety is beginning to have an impact. In June 1973 Dixy Lee Ray, AEC chairman, agreed not to appeal a federal court decision forcing the AEC to make public a detailed environmental review of a new and controversial type of nuclear reactor. She also announced the retirement of Milton Shaw, director of the AEC's Division of Reactor Development and Technology, who, environmentalists charged, had paid little attention to reactor safety.

Such developments have encouraged citizen groups, and environmentalists to ask more probing questions. Are we guaranteed a thousand generations free of sabotage, hijacking, or acts of God that would loose nuclear materials on man and his environment? Are any of the hundreds of planned nuclear plants to be located in places that will threaten human life? What will eventually hap-

---
* Quoted from *Environmental Action,* November 25, 1972.

pen to low- or high-level radioactive wastes now stored in containers with lifetimes of only hundreds of years, whereas the lifetime of wastes is thousands of years? How harmful is thermal pollution due to inefficient nuclear reactors?

Again and again citizens are assured that a major nuclear accident probably won't happen. Nuclear experts belittle such remote possibilities, but they don't deny that they are, nonetheless, very real.

Alternative Environmental Energy Sources

Continued use of chemical and nuclear energy sources at present rates creates a number of serious problems: depletion of limited domestic and foreign resources of fossil fuels and uranium; outlay of immense amounts of money to foreign lands; massive disturbances of land by mining operations; chemical and radioactive pollution of the environment; and the health and safety hazards connected with some of these energy-producing operations. These problems will remain serious until we institute a more rational energy policy.

Ignoring mechanical energy sources (from springs and flywheels), which have little practical significance, let's consider a fourth group—environmental energy sources. These include solar energy, geothermal energy, and energy due to thermal differences in the ocean and to gravitational forces such as tides and wind and water flow, especially hydroelectric power. These environmental energy sources are, by and large, less polluting than chemical and nuclear sources but do have some undesirable side effects.

Some of these undesirable effects could be lived with. Solar energy farms operating at 5 per cent efficiency would require about a tenth of our total desert space. Undoubtedly, some critics would call this "scenic pollution" and claim that another battle had been lost to technology. They would also oppose further damming of rivers for hydroelectric power plants and would fight any move to build windmills throughout the prairie states. But man can improve as well as destroy the natural environment. As René Dubos points out, windmills have added to the beauty of the Netherlands, which had been a forbidding swamp before man decided to develop it. God made the world, but the Dutch made Holland. If man so desires, he can develop and use environmental energy sources.

Solar Energy

Man looks to the sun for his nourishment and life. The sun furnishes energy to make plants grow through the complex process of photosynthesis. The sun gives light and warmth to man, and he rejoices when the days begin to get longer. Primitive peoples in the Northern Hemisphere celebrated the winter solstice, and this carried over into modern times in the feast of Christmas. Today man is looking at the sun in another way. The vast solar energy that strikes the earth is virtually untapped and has no polluting chemical emissions and produces no radioactive wastes. This virtually inexhaustible energy source will become the focus of energy research in the coming decades. Depending on the amount of financial resources we allot to it, solar energy could supply a sizable portion of our energy needs by the early twenty-first century. If just 1 per cent of the solar energy falling on the Sahara Desert in the year 2000 were converted to electric power, it would supply the world's electric needs for that year.

Solar energy can be used to heat or cool buildings, generate electricity by thermal conversion systems or photoelectric power systems, and produce renewable supplies of clean hydrocarbon fuels from biological materials (bioconversion)—although bioconversion will most likely be neglected because so much of the world's good agricultural land must be utilized for food and fiber production.

Approximately 25 per cent of our nation's current energy consumption is for heating and cooling purposes. Experts estimate that 30 to 50 per cent of U.S. heating and cooling energy requirements could be furnished by solar energy with benefits of fuel savings, reduced pollution, and independence from complex energy transmission and distribution systems. An Atomic Energy Commission task force report admits that no major technical barriers exist to development of practical solar heating and cooling systems.

The same report predicts that solar thermal conversion systems, which collect solar radiation and convert it to electric power and thermal energy, will eventually provide 10 to 20 per cent of electric power requirements alone. There are no fundamental technical limitations to widespread application of these systems, but the high cost of solar collectors is a stumbling block. In a community of 2,000 houses using such systems for both electricity and heat, additional residential costs of from $1,000 to $1,500 over houses using conventional systems are expected at this moment. With the exception of solar collection materials, these systems require no more building material than conventional fossil fuel plants.

Large central power systems require solar collector units covering large surface areas of the earth, yet the total land area needed for furnishing all energy at 20 per cent efficiency by 1985 would be less than that kept idle in the farm-subsidy programs of 1972—60,000 square miles. This is not a great deal when we consider that of the 3.1 million square miles of land in the United States (excluding Hawaii and Alaska) 500,000 square miles are farmland, 180,000 represent the roof area of man-made structures, and 100,000 square miles are desert.

The development of photovoltaic electric power requires advanced technology. Solar power units have been developed at great expense for spacecraft, but terrestrial photovoltaic systems for central station power and rooftop use are not yet practical at competitive prices. Even more remote in time and more prohibitive in cost are space stations (synchronous satellites) to provide unlimited power to earth.

The major obstacle to realizing these dreams is our inability to produce large quantities of photovoltaic arrays at low cost. If cadmium sulfide cells are to be used it would take more than the total 1971 U.S. cadmium reserves to generate 1 per cent of U.S. electric power in the year 2000.

More solar energy research is needed. Unanswered questions remain, but not as many as nuclear power proponents would have us believe. Of 4,400 U.S. energy research projects in the mill in 1972 (about $1 billion worth) only seventy have any relation to solar energy. Most were devoted to locating and developing new sources of fossil and uranium fuel. Our annual budget for solar development before the oil embargo was a mere $15 million, whereas we spend more than half a billion annually on nuclear research. It is time to re-examine priorities.

Thermal Energy Sources

Geothermal energy looms as a significant source of electrical energy. Hot springs and geysers exist where faults in the earth's crust allow molten rock to heat underground water. Anyone watching the periodic eruptions of Old Faithful at Yellowstone National Park knows that a lot of steam comes up with each belch. Engineers in the early years of this century devised ways of harnessing the energy in these outlets of steam. At one such site the Pacific Gas and Electric Company uses dry steam with no accompanying moisture to generate about 290,000 kilowatts of power at costs less than fossil and nuclear power plants.

Geothermal plants are in operation in such diverse places as Lardarello in Italy (an early center for geothermal electricity generation), Namaskard in Iceland, Pathé in Mexico, Wairekei in New Zealand, and Matsukawa and Otake in Japan. The Russians have developed a power plant in which the heat from wet steam produces a working fluid of freon for operating a closed-cycle organic vapor turbine. Some energy-poor countries such as Kenya and Ethiopia perhaps could meet growing energy demands by tapping their geothermal fields. In the United States most geothermal fields are in the West, where hot rock can be fissured and used as heat to obtain high-temperature water to run power plants. Potentially, hot rocks could furnish the world with 32 billion kilowatts of electricity, and volcanoes could supply it with another 300 million.

Geothermal energy generation does not come without some cost to the environment. Minerals in the condensed steam can pollute rivers; a characteristic smell of rotten eggs hangs over geothermal sites; the venting of the steam is quite noisy; and land subsidence, which can render structures uninhabitable, occurs in steam-field areas. With care, these costs to the environment can be minimized. The major problem—disposal of high-mineral-content emissions—has been partly solved by returning the condensed brine to underground reservoirs. Allen Van Huisen of Geo-Energy Systems has applied new breakthroughs in heat-exchange technology and has patented a system that uses geothermal steam without displacement of underground water, steam, or brine.

The thermal gradients of the ocean are a second thermal energy source, still in an elementary stage of development. There are places in the ocean where, due to current temperature differences, heat gradients of twenty degrees or more exist between cool- and warm-water zones. A pilot power plant to utilize these differences was tested off the Ivory Coast near Abidjan but was later abandoned. The possibility of utilizing the temperature differences between atmosphere and ocean temperatures in energy-short polar regions has also been considered.

Wind and Water Flow Sources

At one time or another most people have undoubtedly experienced the power of the wind or been swept off their feet by an incoming wave. For years man has tried with varying degrees of success to harness this environmental power. Windpower has been harnessed to grind wheat into flour and to saw timber in Holland and Spain. Winds, waves, and convection and other currents on this earth contain a total of about 370 billion kilowatts of power —approximately ten times the power of geothermal energy sources. Some experts think the wind is the untapped resource most worth harnessing in the next few decades. Homes on the Great Plains and in certain mountain regions, if equipped with windmills, could easily generate their own electricity and store surplus energy in batteries for use on calm days.

The tides have fascinated man from the earliest times. A tidal project off the coast of the United States was considered as early as 1940 but was never completed. One tidal power plant is operating off the channel coast of France today, but there are only a few places in the world where tides could be effectively utilized, and the total power locked in this source is only about 1 per cent of that in the winds.

A more familiar source of water flow power is the flowing river. Many of our great rivers now have massive concrete dams, like the Grand Coulee Dam on the Columbia River and the Hoover Dam on the Colorado. Until recently these were the mainstay of our electrical energy pool. In 1950 hydroelectric power supplied 29 per cent of all electricity used in the United States. By 1972 the proportion had dropped to 15.6 per cent, or 4 per cent of the total U.S. energy supply. Most choice sites have already been dammed —environmentalists are actively attempting to preserve the remaining sites—so the U.S. hydroelectric power capacity can be expected to grow at an annual rate of only 1.6 percent for the next decade.

Hydroelectric power plants cause no chemical pollution, but attempts to expand this source of energy are staunchly opposed—not without reason. Such scenic rivers as the Snake River of Idaho would be turned into a man-made lake for the sake of electric power, and the natural habitats of wildlife in the river basins and neighboring valleys would be ruined. Another drawback is that the lakes formed by the dams collect the silt that is often needed for downstream basins.

Related to hydroelectric power are the newly developed pump-storage procedures, in which nuclear power plants serve as a primary energy source and in off-peak hours pump water into storage reservoirs, to be released as hydroelectric power at peak times. Aside from the danger from nuclear plants and by-products, there is added concern about rapidly fluctuating storage lakes. They would be both a danger to man and unfit for fish and other wildlife.

The Pollution Crisis

The increasing use of fuels has led to a multitude of environmental pollution problems. Smog-filled Los Angeles, filth-laden rivers, and stripped Appalachian countrysides all testify to a damaged environment. Although man has mined coal for 800 years, half of all coal dug from the earth has been mined in the last three decades, and there have been estimates that by 1980 half of all the coal mined in the last 800 years will have been mined in the 1970s. Ever greater uses of fossil fuel overload the capacity of the earth to sustain and neutralize the emissions that result from its combustion.

The dimensions of the energy crisis are a matter of dispute, but few deny that the dimensions of the pollution crisis make it a problem deserving of national attention. We are all aware of the effects of sewage, litter, junk, and smog on plant, animal, and human health. However, people in the fossil fuel and nuclear energy industry point out that man-made pollution is minor compared to natural emissions from volcanoes, dust storms, and evergreen forests. They claim that some pollution is necessary for upholding our standards of living, that excess pollution is being controlled through existing legislation, and that the problems are not as serious as environmentalists suggest.

An instance of this deliberate disparagement of pollution effects is an article that appeared in a 1972 issue of Exxon Oil Company's *The Humble Way*. Its message was: Don't be alarmed at the prospect of increased ocean shipments of petroleum. The article attested to the common occurrence of natural oil spills and was embellished with pseudoscientific arguments and a colored photo showing the merry bubbling of natural gas in the ocean off the Louisiana coast. At a workshop on the effects of petroleum on the marine environment held the following year by the National Academy of Sciences, Big Oil participants disputed many of the data presented, claiming that estimates of artificial oil spills were high and those of natural seepages far too low. Many of the scientists present simply laughed at this propaganda approach to the problem.

Each time major environmental legislation is scheduled to come before Congress for a vote, the major oil and motor companies flood the papers with advertisements stating their views. In 1973 Chrysler Corporation, faced with stiffer emission standards, said in ads that the air-pollution problem was on the way to being licked. No doubt this was true, for man-made carbon monoxide and sulfur dioxide levels in American cities were beginning to decrease. But this was due to the stiff enforcement of the Clean Air Act—an Act that had been opposed by both auto and oil companies.

Less forgivable than industry's attempts to deprecate pollution problems are claims that no pollution problems exist. An example of this approach appeared in a 1973 issue of *Ethyl Digest,* published by the largest lead gasoline antiknock manufacturer in the world, in which time and again the magazine editors openly attack the Environmental Protection Agency, stating that it is manned by "overly zealous environmentalists" and misguided scientists who imagine that lead dispersed in the air and the dust from auto exhausts is toxic. The Ethyl Corporation organ ignores the growing body of evidence showing high concentrations in the atmosphere of air- and dust-borne lead and contending that high lead levels in the blood of some urban children may be traced to automotive sources. The source of a victim's high lead levels is admittedly difficult to prove, but no more difficult than proving that lead from automobile exhausts is harmless.

No one really knows just how much pollution really costs. How does one determine the cost of a death, a major illness? Economic calculations cannot really come to grips with subtle damages to man and his environment. However, our society demands cost-and-benefit analyses. It wants a dollar tag on the national costs of pollution damage. The Environmental Protection Agency attempted such a calculation for a single year, with the results shown in Table VI. The first half of the table considers type of pollutant, the second the source. These calculations are rough approximations only and tend to be on the conservative side. Numerous pollutants and side effects are ignored, such as animal health effects, litigation costs, and addition of antioxidants to rubber due to emissions from autos.

In spite of the tremendous costs to the nation traceable to pollutants, foot-dragging by producers and apathy by consumers have permitted continued profits from polluting sources to pile up. Threats to close a factory or increase prices due to pollution abatement costs are sophisticated forms of blackmail. Threats to the pocketbook and payroll speak louder than clean air and water. Still, it is unfair to gauge the public's willingness to pay for cleaning up the environment by asking such a question as: "How much are you willing to pay for clean air?" A more appropriate question would be: "How much should the polluter be made to pay—his excess profits, his entire profits, or some of his capital investments?"

After a major mercury-poisoning scare in mid-1970, the Environmental Protection Agency focused its attention on chloralkali plants throughout the country. In older plants, mercury cells were leaking large amounts of mercury into the air and waterways, where the pollutant eventually settled in rich organic mud deposits. There the mercury was transformed into more toxic compounds that entered the food chain and reached man through the fish he

TABLE VI
Costs of Pollution Damage in the United States, 1968

BY POLLUTANTS

	Sulfur Oxides	Particulates	Nitrogen Oxides	Total
		In billions of dollars		
Residential property	2.808	2.392	–	5.200
Materials[a]	2.202	0.691	1.127	4.752
Health	3.272	2.788	–	6.060
Vegetation[b]	0.013	0.007	0.060	0.120
Total	8.295	5.878	1.187	16.132

BY SOURCES

	Fuel Combustion	Industrial Processes	Transportation	Solid Wastes	Miscellaneous	Total
			In billions of dollars			
Residential property	2.802	1.248	0.156	0.104	0.884	5.200
Materials[a]	1.853	0.808	1.093	0.143	0.855	4.752
Health	3.281	1.458	0.197	0.119	1.005	6.060
Vegetation[b]	0.047	0.020	0.028	0.004	0.021	0.120
Total	7.983	3.534	1.474	0.370	2.765	16.132

[a] Degradation of materials.
[b] Damage to vegetation and agricultural productivity.

Source: Larry B. Barett and Thomas E. Waddell, *Cost of Air Pollution Damage: A Status Report,* U.S. Environmental Protection Agency, Research Triangle Park, North Carolina. Estimates were not calculated for soiling and aesthetic effects in order to avoid counting twice those damages affecting property values.

ate. Olin Mathieson Corporation, realizing that their chloralkali plant at Saltsville, Virginia, was a major polluter, simply closed the plant rather than clean it up. They left behind empty factories and polluted streams, mute testimony to corporate irresponsibility. Similar cases would be fewer if polluter executives were hauled into court and sentenced to put on boots and clean up polluted streams themselves.

Modern technology has been quite innovative in spawning new pollutants along with new products. The average citizen is not anxious to hear about these new dangers and is generally quite bewildered by the tremendous variety already brought to his attention. He doesn't want to be told that spreading sewage on farmlands may cause secondary pollution effects, or that the large quantities of trace metals now finding their way into urban dust due to increased fuel consumption may, like cigarettes, "be harmful to your health." Citizens can become conditioned to filth. And they can learn to live with danger, much as the Maltese citizens did in World War II while surviving a thousand German air raids. But even such conditioning qualifies as an uncomputed damage from pollutants.

The Crisis in Technological Priorities

Environmental energy sources offer immense promise, but their development presents problems. The trillion-dollar question is whether the people of the United States are willing to review present commitments and set new priorities. Presently, about 46 per cent of our national budget is for military activities, veteran benefits, and space exploration. Pressing demands on the remainder of the budget to fill immediate needs for federal social programs make it difficult to scrape together the vast amounts of money needed for an effective program for alternative energy sources.

Any discussion of development of energy sources requires a division into medium- and long-range programs. Such long-range programs as solar energy development will cost billions of dollars —a commitment equal to or greater than that of the space program. The recently increased but still modest research-and-development energy funds are also being sought after by fossil fuel producers. Granted that coal gasification and liquification and exotic electric generation technologies (like magnetohydrodynamics) do offer promise for more efficient use of our dwindling fossil fuel pool, one must face the fact that efficiency is often *not* profitable so long as an energy policy permits depletion allowances and makes freight rates conducive to the use of cheap stripped coal. Contrary to popular opinion, our power plants have become less—not more—efficient in recent years. Joseph Swidler, Chairman of the New York State Public Service Commission, has said that in 1962 it took 8,588 British thermal units of fuel to produce one kilowatt hour of electricity; by 1971 the most efficient plant required 8,695 Btu's per kilowatt hour. Averages are even more revealing. This growing inefficiency may be partly due to use of gas turbines. Just to return to pre-1965 techniques would save 0.4 million barrels of oil per day. Figures 1 and 2 in Chapter 2 graphically demonstrate our growing inefficiency in the use of energy. (It should be noted in passing that electric power companies have one of the poorest research-and-development records of any major industry class.)

Short-term efficiency techniques are economically costly, but they do result in savings of natural resources. The major problem here is developing long-range energy alternatives of an environmental nature. To establish these new technological priorities will require proper technology assessment (see Chapter 4), a choice of viable alternatives in an integrated energy and natural resource policy (Chapter 5), and a firm commitment for funds to carry out the needed programs (Chapter 6). Few businessmen champion long-term environmental energy programs since profits are uncertain and the investments risky. Energy priorities are thus not set by actual need but by the short-term profitability of the oil and coal industry.

Potential long-range energy sources such as nuclear fusion present a number of technical problems. While fusion sources appear inexhaustible, the feasibility of nuclear fusion has not yet been demonstrated. Scientists are worried about possible radioactive tritium, lithium, and beryllium by-products. When health and safety hazards are linked to long-term investment in an energy source, the source might just as well be counted out for consideration in this century.

A willingness to entertain shifts in technological priorities requires a clear perception of the energy problem, a consideration of alternatives to present practices, a rational discussion of necessary steps to make these operative, and confidence that the change is justified. Perhaps the heart of the American crisis in will power is that we are afraid to change and fearful of what the future will bring. We raise objections to the need to change or find distractions to turn us from the immediate problem. We deny any energy crisis, preferring to attack less expensive problems. Meanwhile, the masters of commercialism plead: "Trust us; we understand this energy problem. We can keep you comfortable. Drive a little slower, keep the car in shape, turn the thermostat down (or up), and leave the rest to us."

In times of war and famine, citizens frequently turn to dictators and authority figures. Such an alternative might appear tempting to people living in countries experiencing severe materials shortages. But strong-arm tactics are no substitute for developing the will power to conserve our resources.

The so-called Club of Rome, in *The Limits of Growth,* predicts the collapse of society through depletion of natural resources due to human shortsightedness. In classic economics the Ricardian view is that for every scarce material a substitute can be found, though at a higher cost. Americans tend to go along with this, putting a blind faith in technology's power to solve each short-lived crisis by providing new technology. Sir Charles Darwin, in his 1953 volume *The Next Million Years,* contends, logically enough, that the Ricardian view depends for its validity upon an infinite energy source. Perhaps we might extend his observation to include an infinite financial and psychic energy source, which implies the ability to buy anything and the psychic nerve to live through any crisis. We have neither, so it is better to develop some realistic strategies for meeting current crises.

Many of us are familiar with the Biblical parable about the ten bridesmaids who brought lamps to meet the bridegroom. Half of them brought sufficient oil and half did not. In the middle of their wait for the bridegroom, the foolish bridesmaids observed that their lamps were going out and asked the wise ones for more oil. But the wise bridesmaids refused, saying that there would not be enough oil for both themselves and the foolish bridesmaids.

The world has both wise and foolish users of oil. Some, like the Chinese, have learned to conserve their limited fuel; others, like the North Americans, have never learned to conserve. They foolishly squander their natural resources, and when their energy stores run low, say to the Arabs—or Indonesians or Nigerians—"Give us some of your oil."

Americans have added a new twist to the parable. They have decided to wait for the bridegroom (Technology) in the hope that he will come to their rescue and convert environmental energy sources (the tide and the wind) into usable energy. But financial difficulties have caused a delay in the bridegroom's arrival.

Perhaps our combined crises can be summed up as a lack of willingness to face up to the shortages that are before us. We want to retain our outmoded standards of living and our profit-motivated economy, which are geared to using as much energy and materials as necessary to keep them alive. But to continue our past energy practices in the face of growing world shortages is foolishness. And we must choose between foolishness and wisdom.

4 · Assessing Our Needs

For a brief period of human civilization man lit up this dark earth with the hydrocarbons that had been stored for millions of years under its surface. If we diagram this all too brief period of fossil fuel consumption over the centuries, the result will be a "fossil candle," peaking sharply toward the end of the twentieth century (Figure 4).

In the early 1970s 95 per cent of our energy was derived from fossil fuels (oil, 46 per cent; natural gas, 31 per cent; coal, 18 per cent), whereas hydropower, an environmental energy source,

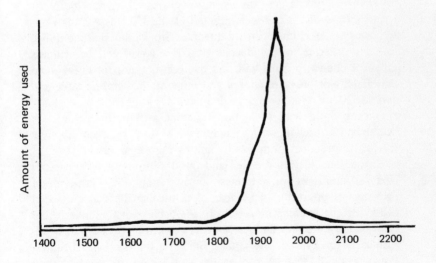

FIGURE 4. The "fossil candle."

accounted for only 4 per cent and nuclear power for about 1 per cent. Macbeth's cry of despair "Out, out, brief candle!" applies only too well to this fossil candle, which is flaring brightly at the present time but whose coming extinction gives us cause to reflect.

Sensible individuals prepare for the future through insurance policies, retirement funds, and savings. However, these preparations are usually of benefit only during one's own lifetime. Future generations are not considered. As one oil company's nuclear expert said, in addressing an audience that objected to the generation of nuclear wastes and their disposal in containers that may not last beyond a century, "That's another generation's problem. We'll be dead and gone by then."

The specter of immediate and future energy shortages and problems is sobering. It requires that we plan ahead. Philosophically speaking, this implies gathering data, determining which problems to tackle first, developing alternatives to outmoded practices, and weighing dispassionately the advantages of these alternatives through careful assessment.

Environmental planning is a global exercise, demanding collaboration of men of good will. It involves assembling facts about the extent of ecological damage, sorting pollution problems by order of magnitude, and discussing ways of curbing specific problems. Such a discussion was initiated at the United Nations Conference on the Human Environment at Stockholm in June 1972, but any exercise in global ecological discussion will be fruitless unless it comes to grips with the overconsumption of energy and materials. Not only are energy and minerals becoming scarce and expensive, but scarcity and high prices are triggering food shortages that could lead to a global famine within this decade. A heightened awareness of these matters is of no consequence, unless it leads to practical strategies for improving the situation.

Throughout history the rich and the unscrupulous have capitalized on shortages and famines. During such times criticism of profiteers is often muted for fear of disturbing the delicate social order. People become more selfish. It's each man for himself and each nation for itself, undoing alliances that took decades to build. France is for France; the United States is for the United States. Power plays appear. A run on the world bank of materials and energy supplies develops, and panic rules the day. To counteract all this, global cooperation during an era of shortages must be universal, prompt, and radical. The need is for decisive action and sacrifice. If some are allowed to profit when others are cutting back on production and consumption, then the whole global social order is threatened.

A Need to Listen: The Lesson of Stockholm

Something new was happening in Stockholm on Monday, June 5, 1972, as official and unofficial delegates from Africa, Asia, and the developed countries of the West to the United Nations Conference on the Human Environment assembled for a moment of global reflection. They were aware that the earth's environmental problems could not be solved by unilateral decisions but demanded international cooperation. In her book (*A New Creation? Reflections on the Issue,* published a year after the conference, Barbara Ward, one of the conference's major architects, had this to say:

> The events of Stockholm did not seem routine. They had an aura of significance that far transcended the daily transactions. People wondered whether they were witnessing a turning point in history, an end of an era, the beginning of a new age. The mood was not discouraged. Rather, it was wary. Yet the question was being put: have we seen the beginning of the end of a society built by post-Renaissance man?

This wariness pervaded both official delegates and representatives of nongovernmental organizations; it permeated the press and filtered down to the ordinary citizens who stopped by to see exhibits and hear talks at the half dozen centers set up for environmental information. Technocrats from developed countries saw their pet environmental solutions shot down by Third World radicals. The idealists' only-one-world theme proved more dream than reality, as philosophical differences shattered expectations of easy solutions. Russia and its satellites had chosen not to attend, since East Germany was not represented. The Vietnam War was labeled a "grand distraction" and precipitated numerous demonstrations and talk of ecocide.

The mood in otherwise placid Stockholm was also one of wariness, as literally thousands of troops lined the streets, anticipating a horde of hippies who didn't come. The U.S. delegation, stung by antiwar demonstrations and Administration strictures on what they could and could not do, withdrew into their hotel rooms. In conference voting, they often found themselves in the minority—55 to 2 and 57 to 1.

The conference was without "leaders." The developed nations were the foremost polluters, and the Third World representatives were divided among those wishing to mimic the West and those desiring radical change. The Chinese made a power play to gain leadership but refused to accept the set of basic principles drawn up by a planning committee. Their insistence on speaking Chinese, an official United Nations language, caused a temporary translation crisis, and for a critical day or so, the other delegates simply didn't know what the Chinese were saying.

Idealism clashed with realism on the streets and in the conference halls. A parade of Japanese victims of crippling mercury effluents from industrial plants on Minamata Bay made these poor fishermen look like wounded sheep in the eyes of an unbelieving world. Displays of photos of Vietnam War victims and scarred landscapes added to the horror of an already tragic war.

Amid this dramatization of what man was doing to his fellow man, there was gradually hammered out over a two-week period a *Declaration of Principles,* which laid the groundwork for an environmental ethics. The principles were supplemented by the so-called Stockholm Action Plan—109 resolutions that were little more than expressions of good will. Some were doomed before the ink dried. A ten-year ban on whaling was included but was disregarded a few months later by the principal whaling nations, Japan and Russia. A resolution to halt French nuclear testing in the South Pacific was flouted even more flagrantly. Delegates from New Zealand and Peru, plus a demonstration by the Green Peace Movement, failed to halt the testing, and the bomb went off while the conference was still in session. However, the United Nations did endorse the Stockholm Action Plan and set up a new headquarters in Nairobi, Kenya, in 1973 for its implementation.

Conference participants learned quite quickly that environmental problems require more than pious resolutions. The world's most powerful nations were challenged on certain issues: the United States for its Vietnam policy, France for its nuclear testing, and Japan and Russia for killing off the whale population.

At Stockholm, amid posters, talks, demonstrations, and movies of man's lack of respect for man and nature, a consciousness grew that this problem was quite complex and required solutions on a global scale. Man was steeped in environmental sin, and the Stockholm Conference was the confessional. The world was outgrowing nationalism, yet nationalism was evident in many of the conference procedures. It became apparent that United Nations conferences are incapable of dealing with commercial exploitation by developed countries. As long as there is no police and regulatory body to guard the ocean and atmosphere, pollution is bound to go unchecked. Unfortunately, both developed and developing nations seemed to spend more time defining national than international rights.

The participants at the conference worked under a number of handicaps. The Third World found that it had little direct experience of either overconsumption or pollution. The forty nations of Africa have collectively less auto pollution than the city of Los Angeles. Likewise the West was working under a handicap of not having developed a clear-cut plan to modify consumption and simply refused to re-examine its "cultural necessities," its entrenched industrial practices, and its continuing violence to forests, rivers, and air.

The atmosphere of the governmental and nongovernmental forums at Stockholm was not conducive to fruitful communication, but at times symbolic action had more impact than words. Third World radicals took over a panel composed mostly of white Western environmentalists. They shouted down Paul Ehrlich, author of *The Population Bomb,* and demanded equal time for presenting their own contributions. While this irritated many of the predominantly white audience, it symbolized the struggle of the Third World to gain a hearing for its positions and proposals on an equitable footing with that of the West.

Some Westerners found the experience quite educational and awakening, having come to Stockholm with the naive assumption that developed nations had the only solutions to the environmental crisis. One group of sympathetic allies of the Third World radicals was the Holy See Delegation, which delivered a message from Pope Paul:

> No one has the right to take over the environment in an absolute and egoistic way. The world man lives in is not a *res nullius,* the property of no one; it is a *res omnium,* the patrimony of mankind. Those responsible for the environment, both private and public agencies, must regulate the environment for the well-being of all men, for man himself is the first and the greatest wealth of the earth.

The earth, in short, belongs to all of us, and all of us bear a responsibility to see that it is preserved and shared. A billionaire oilman has no more right to call the oil he pumps his, than does the penniless peon who lives beside the pump. *Res omnium* is a Judeo-Christian concept, derived from the belief that God created the earth to be shared by all. Truly there is no no-man's land anywhere on this earth, for the community of man shares in the use of oceans, mountains, and air. We all share in their stewardship; it can't be left to some noble class. Great disparity in wealth and in uses of the world's resources must be done away with if we expect a renewed social and physical environment.

It may be that the Stockholm Conference was the temporal event and Stockholm the physical place where "cosmic" man was born, for it was in Stockholm that many of us first learned to see—and to listen to our fellow man.

Listening is an art. It demands the respect of all present. It makes it possible for all to speak in turn and to learn from the experiences of others. Listening is an essential part of the human dialogue. It is more than just hearing audible sounds; it includes comprehension of what the other party says and a willingness to respond meaningfully; it is one of the prerequisites of an environmental philosophy.

The Need for a New Philosophy

> Man has the fundamental right to freedom, equality and adequate conditions of life, in an environment of a quality which permits a life of dignity and well-being, and bears a solemn responsibility to protect and improve the environment for

present and future generations. In this respect, policies promoting or perpetuating apartheid, racial segregation, discrimination, colonial and other forms of oppression and foreign domination stand condemned and must be eliminated.

Principle 1, Declaration of Principles, United Nations Conference on the Human Environment

The first of the twenty-six Principles of the Declaration formulated by the United Nations Conference on the Human Environment states that the basis for environmental justice is a balance between human rights and responsibilities. Imbalances such as colonial oppression must be faced and exposed for what they are before we can conserve and rebuild the earth. A balanced social order is a precondition for improvement of the physical environment, for man and his world are one. That is the substance of *Principle 1*.

The subsequent principles spell out some of the consequences of the first principle. The natural resources of the earth must be safeguarded for the benefit of present and future generations. Those who live for today only, knowing that tomorrow they die, may be technologically sophisticated, but they are ecologically barbaric. Appropriate planning and management includes maintaining, restoring, and improving the capacity of the earth to produce renewable resources. Wildlife must be protected and nonrenewable resources used wisely. (*Principles 2–5*)

The poacher and the fur seller may deserve blame for the demise of certain species, but so do the buyers of furs and the members of a democratic society who permit such sales to occur. We can't allow our people to discharge toxic substances in such quantities as to overload the earth's capacity to detoxify them. It is not enough to conserve, preserve, and improve the environment; we must foresee where and how we can harm or destroy any part of the ecosystem and refrain from these practices. (*Principle 6*)

One part of the ecosystem singled out for special mention is the oceans. The immense increase in oil tanker traffic flaunts the principle that says states shall take all possible steps to prevent pollution of the seas. (*Principle 7*) Few steps have been taken jointly as yet, and the pollutants are increasing to such an extent that even the oceans are not capable of detoxifying the polluting substances through natural processes. The oceans may belong to all, but they are policed by no one. The illegitimate uses of them are bound to continue.

We might summarize and draw a corollary from the first seven principles:

> *In order to share these rights and exercise these responsibilities, man must unite in social units capable of carrying out his ecological duties. Environmental control and protection is beyond the capacities of individuals; it requires social structures.*

Man, being both economic and social, must be assured of good working conditions for development. This requires social regulation. Furthermore, in considering the grave problems arising from deficiencies due to underdevelopment and natural disasters, the Declaration states that such deficiencies "can be remedied by accelerated development through the transfer of substantial quantities of financial and technological assistance as a supplement to the domestic effort of the developing countries." (*Principles 8–9*)

Collective responsibility means that the earth's people must *share* in each other's development. Sharing is a surrender of one's individual or vested interests for a greater good. It involves such economic factors as stability of prices (and adequate earnings for workers) in primary commodities and raw materials and the economic health of all nations, including the developing ones. Environmental measures taken by developed countries must not adversely affect the developing nations. Since remedial environmental measures cost money, many poorer countries do not have the resources to pay for them, and here international technical and financial sharing is necessary. (*Principles 10–12*)

An additional aspect of sharing, omitted from the Declaration, stems from insights obtained at some of the meetings of nongovernmental organizations at the conference, concerning the need of developed countries to acquire some of the qualities of life preserved more or less intact in many of the developing nations. Developed countries must be willing to receive as well as to give, to be educated as well as to educate. The East needs material development, but it can also provide the West with examples of ascetic practices needed for a balanced global ecology. The discipline of the Chinese, the respect for nature of the Southeast Asian, and the social community sense of the African are positive contributions of as great a value as antipollution techniques and devices.

The sharing of resources (whether cultural or material) will occur only when all nations undertake programs of preparing their people for the task of global environmental cooperation. Cooperation demands that all citizens know the facts, so that they can help formulate and make proper judgments for rational management of natural resources. To ensure the basic freedoms enumerated in Principle 1, (freedom, equality, and adequate conditions of life), global education on a level not previously recognized is necessary. Thus, a second corollary follows:

Environmental cooperation can be effectively exercised only when citizens have been elevated to a literate state. Global environmental education is the task of all nations but rests more heavily on developed countries responsible for environmental problems.

The Declaration advocates an integrated and coordinated approach to development planning, especially when there is a conflict between the needs of development and the need to protect the environment. Many developing countries are plagued with unplanned urban growth far beyond what was expected just a decade ago. These conditions breed crime and slums. Development planning must therefore be applied to all human settlements, to urbanization, and to demographic policies.* (*Principles 13–16*)

---

* The conference deferred consideration of population problems to the 1974 World Population Year Conference in Bucharest. The working principles of that Conference include:
• The sovereign right of each nation to decide, on the basis of informed awareness, what its own population policy should be, is respected.
• The personal right of a couple to decide the size of its family is respected.
• The social and cultural value differences among nations and communities are recognized in formulating and implementing population programs, most of which are, therefore, country-based.
• The world population situation is regarded as a complex matter which varies in its nature and substance from place to place rather than as one responsive to a simplistic over-all solution.

The Stockholm Conference recognized that planning, managing, and controlling environmental programs need appropriate institutions for their administration. Such institutions would muster the scientific and technological resources for the identification, avoidance, and control of environmental risks and for the common good of mankind. (*Principles 17–18*)

The above principles are necessary prerequisites for an atmosphere conducive to environmental education. The Declaration states that such education is essential "in order to broaden the basis for an enlightened opinion and responsible conduct by individuals, enterprises and communities in protecting and improving the environment in its full human dimension." (*Principle 19*)

Without mastery of the written word ecological education is nearly impossible. Ecology is a complex subject and its understanding demands literate citizens. Notwithstanding herculean efforts on the part of many developing nations, about a third of the people of the world are either actually or functionally illiterate. Some of the affluent among us might conceivably approve of this situation, claiming that literacy campaigns cause an undercurrent of discontent among "have-nots" and are economically costly. The possibility of discontent can be neither denied nor condemned. Still, the expense of a global literacy campaign is often exaggerated. Experts tell us that the cost would be about eight dollars per person in developing countries (or $8 billion to affect a billion people). Spread over a five-year period such a program would cost $1.6 billion per year, or about half of U.S. overseas aid each year. Not everything costs as much as it does in the United States. Some developing countries are building mass housing for as little as $10 per unit. Literacy, like adequate housing, safe drinking water, and basic hygienic systems, is not a high-priced luxury.

Principle 19 goes on to say that the mass media must not contribute to the deterioration of the physical or cultural environment. The mass media is a major pedagogical tool of our age and, as such, should play a role in environmental education. Each American is subjected to a bombardment of commercials, costing about $20 billion per year (or about $100 per person), on the airwaves. Television and radio are now becoming available in every part of the world, and so are commercials. When certain countries forbid the airing of commercials, commercials creep into towns and villages on the movie screens, and pressures to consume more goods mount in the homes of simple people. Consumption increases, production goes up to meet new demands, and pollution and depletion of resources ensue.

Environmental education will have to be conducted in part in the mass media, which reaches the bulk of the adult population. Communications media will therefore have to be free. Thomas Jefferson once said that he would prefer a free press with no government to a government without a free press. Our airwaves, controlled by commercial interests for all practical purposes, are free in name only. Anticommercials, which tell why many particular products are unsafe, unnecessary, or worthless, are being produced for radio and television, but it is virtually impossible to broadcast such messages on prime time. Only a few stations in this land of the free devote time to these countercommercials—but their number is gradually rising.

Proper environmental education includes presenting both sides of controversial issues in such a manner that commercial interests do not predominate. One well-placed advertisement in a newspaper like the *New York Times* or the *Washington Post* can swing an environmental vote in Congress. Quite often only the big industrial interests have money available for buying such ad space at the precise time when it can be highly influential.

It is also important that scientific research and development be promoted in all countries, especially in the developing nations. (*Principle 20*) We cannot solve major environmental problems by merely supplying citizens with information; we have to have the scientific and technical tools to clean up the environment and to control polluting sources.

A third corollary follows from Principles 13–20:

> *Global environmental awareness demands mechanisms whereby citizens can come to a proper and reasonable assessment of the technology which threatens to cause increased pollution or depletion of resources.*

The earth belongs to all, not just to a technological aristocracy. Yet without technical help the average citizen cannot address modern technological issues. However, technological decision-making must be decentralized, and it must incorporate the citizens' views. Many of the voiceless of the world are becoming skeptical of the "trickle-down" theory of Panglossian economics, which says that if the rich build new houses this will benefit the poor, because the houses at the other end of the scale will somehow trickle down to them. In this day and age, when it is possible for men with technical expertise to decide the fate of the earth in corporation back rooms, it is more necessary than ever that the average citizen, who previously saw little need for such participation, take part in the decision-making.

A comprehensive environmental philosophy must spell out the relationship between nation states and the global community of man. The Declaration of Principles states that there is a sovereign right on the part of states to exploit their own resources pursuant to their own environmental policies. (*Principle 21*) However, with this right goes a responsibility not to cause damage to the environments of other states. Switzerland, generally considered to have enlightened environmental policies, had a practice—until it was publicly exposed—of dumping industrial wastes on neighboring nations. Some of the country's most polluting industries are located at Basel, where the waters of the Rhine leave Swiss soil to flow into the German Rhineland and French Alsace.

Cooperation among nations must include payment of liability and compensation, respect for the values of other nations, and equal stature for all nations irrespective of size. International organizations must be allowed to play a vital role in the protection and improvement of the environment. Finally, the states should eliminate and destroy nuclear weapons in the name of a better human environment. (*Principles 22–26*)

These twenty-six Principles, with all their limitations, form a basis for a new environmental philosophy. If the ideas they express are permitted to be forgotten, the Stockholm Conference will become just another page in the history of man struggling to collaborate with man. If they are not to be condemned to obscurity, they must have exposure, discussion, and publicity. Otherwise they remain purely an academic exercise.

The twenty-six Principles point to an environmental thermodynamics, based on the law of conservation of energy. To preserve, protect, and develop this earth takes energy, and there is only so much of it available in the closed system called "Spaceship Earth." The sources of energy are both limited and unevenly distributed. In order that man preserve a continually threatened environment, it is necessary for him to cooperate with men from other places in the use of energy. Global cooperation is good environmentalism.

In the remaining pages of this chapter, we will consider what Americans must do to implement the ideas expressed in the three corollaries derived from the twenty-six Principles just discussed. We tend to think that the development of social structures, access to environmental information, and mechanisms for making technological assessments are problems only for developing nations. This is not true! We have a long way to go in this country. We are in dire need of a structure for weighing environmental alternatives; we must break down the barriers imposed by trade secrecy in order to get the environmental information necessary to arrive at prudent decisions; and we must see that Congress's Office of Technology Assessment is operated in such a manner that citizens can participate in making decisions about how our resources are to be used.

A Need to Weigh Alternatives

Quite often several possible courses of action will lead to immensely different environmental consequences. Without the ability to foresee exactly what these consequences will be, we are still bound to consider alternative courses of action. In formulating a national energy policy, one course of action might bankrupt our country (by increasing imports); another might destroy our landscape (by increasing development of domestic fossil fuel sources), and still another might demand a radical change of lifestyle (by cutting back consumption of energy during the next decade).

A major report of the National Petroleum Council, *U.S. Energy Outlook,* issued in December 1972, presented the three above-mentioned alternatives but dismissed two of them in less than a page and proceeded for over 300 pages to talk about domestic energy development. This is hardly weighing alternatives.

After the Arab oil embargo experience, Americans of many

political persuasions were able to see numerous reasons for not being too highly dependent on oil imports. Environmentalists cited continued depletion of the world's energy resources, increased ocean pollution, and the danger of disasters at deep seaports and terminal sites. The military-minded complained of overdependence on foreign energy supplies in times of national emergency. Economists spoke of increased oil prices and the consequent dollar drain. However, a reasonable discussion of alternatives might uncover some advantages to a policy of increased oil imports that are often overlooked: Increased oil imports might encourage more trade and economic stability in various parts of the world; dependence on fuel might encourage interdependence among nations; the over-all environmental impact of this policy might prove less destructive than if we increased domestic energy production from nuclear and fossil fuels. The issue is not whether these points have greater or less merit but that they are worth weighing when we talk about a national energy policy.

The third alternative (restriction of energy consumption during the next decade) also received short shrift in the National Petroleum Council's report. The reasons given for this curt dismissal included the adverse effect of alteration of lifestyles on employment, economic growth, and consumer choice:

> Despite possibilities of extreme changes or revisions in existing social, political and economic institutions, substantial changes in lifestyle between now and 1985 are precluded by existing mores and habits, and by enormous difficulties of changing the existing energy consumption system.

The report added that efficiency improvements may be desirable but need long lead times and are expensive. What was omitted was that this alternative would not be overly profitable for the fossil fuel producers.

A reasonable weighing of alternatives requires that we ask some questions about restricting energy growth: Could a lifestyle change be simpler and yet for the better? Could a reduced standard of living be good? Could we hold energy consumption at 1972 levels by a series of immediate, medium-range, and long-range strategies? (Immediate conservation measures did work during the oil embargo and resulted in better than 10 per cent savings in energy through March 1974.)

Lifestyles requiring less than half the average American's use of energy are common in Western Europe. Such changes would require less mining of coal, less stripping of our lands, less drilling, less oil spills, less auto use, and less appliance use. This would mean less air pollution, less money for environmental clean-up, less world competition for resources, fewer international tensions, and, in the long run, less need for military strength—and a better world. More use of energy does not automatically mean a better quality of life, even though it might help to quantify a standard of living.

A high quality of life implies a variety of lifestyles to choose from, an ability to move from one lifestyle to another, and a respect for the other person's lifestyle—provided it does not harm or hinder someone else's. One American may seek a simpler lifestyle and move to the hills, grow his own food, and stay there the rest of his life. Another may desire mobility, cultural advantages, and variety in foods and consumer products. "Quality of life" should include both free choice and respect for one's neighbor.

The standard of living of the rich may be high and the lifestyle different, but the conditions are not qualitatively good. A wasteful use of nonrenewable resources by some, while others lack the bare essentials of living, does not contribute to the social environment or the quality of life. Conversely, a heavy use of energy by some, in order that others might have more than the bare essentials and might continue to grow and develop, is contributive to the social order and the quality of life. A standard of living may be high and good as well as high and bad; a lifestyle may be more complex and yet be good by contributing to the social environment.

If we are to entertain changes of lifestyle, some general characteristics should be enumerated:

1. A good lifestyle must tolerate change. It must be open to growth and development. Freedom in choosing alternative lifestyles is paramount, together with freedom to travel and freedom to choose from a variety of products. While both greater mobility of people and availability of consumer goods require transportation

energy, these should be looked upon as part of our exercise of freedom.

2. The lifestyle must use energy efficiently. Note that there is no arbitrary limit placed on how much energy should be used, only that it not be wasted or misused.

3. The lifestyle must not be at the expense of other people. It should use renewable energy whenever possible and leave nonrenewable energy for essential operations.

A lifestyle is both quantitative and qualitative; its efficiency and its use of nonrenewable resources can be measured. Its mode, goals, and world view are qualitative. The Lifestyle Index at the end of this book is concerned with the total energy used by individual Americans. The Energy Units used in the calculations (1 E.U. = 10 kilowatts of electricity) do not take into consideration the fact that 96 per cent of our energy is nonrenewable (including nuclear fuels) and over 50 per cent is wasted through technological inefficiencies.

A lifestyle is dependent upon both social and individual factors, such as how far one lives from work, what means of transportation are available, and how much one is taxed for social and governmental operations. Work attitudes vary with—and help to determine—lifestyle. Some regard work as money-making; others see it as a service undertaken for others but still "gainful" and qualitatively profitable. The work ethic is intertwined with nationalism and patriotism. It is part of U.S. culture, for instance, to speak and think and act in terms of maximization of profits and economic growth.

Part of the resistance to change of lifestyle is the fear that we will have to give up what is good. Some of this resistance has a legitimate basis. Psychologically, to restrict man's innate desire to grow and enrich his life is inhuman. But growth is more than pay raises and bigger homes; it can be measured in qualitative ways. "Go West, young man" and "Blaze new frontiers" are outmoded slogans when there are no new frontiers. Now lifestyles that require economic growth often clash with lifestyles that value open spaces, fresh sea smells, and silent places. No resolution can be achieved unless both types agree that we need a balance of technology and nature, of man's handiwork, and of unmarred beauty.

Assessing Our Needs 81

These differences in viewpoints have much to do with how one understands national goals and determines national policies. The National Petroleum Council, which had previously stated that our country would not tolerate lifestyle changes, listed a number of national energy recommendations in the report cited above:

1. Coordinate energy policies.
2. Establish realistic environmental standards.
3. Establish realistic health and safety standards.
4. Encourage greater development of resources on public lands.
5. Assure water availability for energy production.
6. Continue tax incentives. ("For example, the 1969 Tax Reform Act alone placed an additional tax burden on the domestic petroleum industry of some $500 million per annum.")
7. Maintain oil import quotas.
8. Investigate feasibility and desirability of greater use of electricity generated from domestic coal and uranium resources.
9. Maintain uranium import controls.
10. Allow field prices of natural gas to reach their competitive level.
11. Rely primarily on private enterprise.
12. Expand research in such fields as: exploration methods and equipment; production of synthetic fuels; more efficient production and use of energy; coal mining technology; greater recovery of gas and oil reserves; development of new energy forms.

Growth advocates argue that these proposals will enhance national security, ensure freedom of consumer choice, help mitigate the growing trade deficit, and promote economic growth. From an environmental standpoint some of these (1 and 12) are good proposals, some of them (2, 3, 7, and 9) are good if properly understood, and some (4, 5, 6, 8, 10, and 11) are totally unacceptable. The need for a coordination of a realistic energy policy and expanded research in energy production methods is generally accepted today. However, a realistic environmental, health, and safety policy means different things to different people. A realistic policy to some means guaranteed profits. To others it means the right to fresh air and safe employment practices. To rid our country of black-lung disease is "realistic" to a coal miner but not to some mine company executives. The principle of growth at any cost has made a mockery of free enterprise. Strip mines, radioactive wastes, and exploitation of public lands are part of the price we must pay.

Walter Heller, testifying before the Senate Committee on Interior and Insular Affairs, says that large depletion allowances, capital gains shelters, and special tax deductions should no longer be allowed for the energy-producing industries. He continues:

> Here is another case where the believers in the market-pricing system ought to live by it. The public is subsidizing these industries at least twice—once by rich tax bounties and once by cost-free or below cost discharge of waste and heat.

The pressure applied by Big Oil to shape our national energy policy along nonrenewable energy growth patterns is immense. There are several reasons why development of natural resources to meet present consumption patterns is not wise:

1. Such a program will deplete known energy resources at an ever accelerating rate, and we are already quite wasteful of our resources.

2. Current chemical and nuclear energy sources are highly polluting. Strip-mine and shale oil development are detrimental to the environment.

3. We would have to spend somewhere between $215 and $311 billion in the period of 1971 to 1985 to achieve the wanted growth. An additional $235 billion would be required for power plant construction and transmission facilities and an additional $0.7 to $1.1 billion for water requirements, bringing the total capital needed to $451–$547 billion. Since development funds are finite, this heavy drain in financial reserves means that fewer funds would be allocated to the development of less-polluting environmental energy sources.

4. National stability and defense do not warrant overuse of our current energy reserves. In prolonged periods of national emergencies we might have to draw more heavily on available domestic resources.

For all these reasons, neither increased imports of fuels nor expanded domestic production answers basic problems as to financial stability and environmental preservation. Former Secretary of Interior Stewart Udall says that we will not come to grips with the real issues until we develop policy guidelines that tie us to an ethic of national thrift and gear up a huge research effort that will produce environmentally sound solutions to our long-term energy requirements. He adds that we need in the next decade some drastic but practical reforms that will depress demands that are profligate or unessential.

The United States may be willing to undergo short periods of voluntary energy cuts, but can the people resist for long the constant pressure to consume more and more materials? Without open discussion of the various alternatives, citizens might be in danger of yielding to such pressure.

A Need for Information: The Battle over Trade Secrecy

One cannot successfully weigh alternatives unless one has all the necessary information. Citizen access to relevant information is critical to making proper judgments. In several key policy-making areas the needed information is in the hands of those who already espouse a special interest, and the release of that information is governed by their friends in the federal government. This is true of such areas as the extent of natural gas and oil reserves, substantiation of advertising claims, and the actual amount of pollutants emitted from particular plant sites. Another example of withheld information is the inability of consumers to find out from a manufacturer the nature of complaints about a consumer product. Information is collected but is not made available to citizens. Here legal action is needed to clarify citizen rights to protection through proper health and safety information.

Part of this legal action deals with company "trade secrets." Many consumer products contain chemicals dangerous to man, but when users request information on the chemical composition of such products, the information is withheld, even though the chemicals might be carcinogenic (cancer-causing), mutagenic (causing permanent change in hereditary matter), or teratogenic (causing birth defects). The average citizen has no way of knowing if and how much of these synthetic chemicals either gets into the environment or enters the products he buys. He may want to be cautious but doesn't know where to begin. He may dislike and condemn food additives, aerosols, fluorides, commercial fertilizers, pesticides, household chemicals, and other such materials, but he feels powerless to do anything about them.

This widespread uneasiness about the products of our technology is not a sign of national neurosis or faddism but a perfectly legitimate worry about the violation of a fundamental citizen right: the right to know what materials may harm or threaten one's health and environment. Since consumer products become part of the general environment through use and disposal, this right to know should include information on the type and amount of chemical additives.

The public's right to know is partly protected by the Freedom of Information Act of 1966, which allows citizens access to governmental files and proceedings. This right is honored in the Clean Air Act of 1970 and the Water Pollution Control Act of 1972; it is the basis for disclosure, labeling, and substantiation of advertising by the Food and Drug Administration and the Federal Trade Commission.

Certain rights clash with other long-established rights, and time is required to clarify and evaluate the situation. Environmental rights and the citizen's right to know have come into conflict with the producer's traditional right to withhold proprietary information (known as trade secrets). Producers admit a responsibility to instruct consumers on how to use their products, and they know that selling guns or drugs to minors is an abuse. But they also believe they have a right to keep secret certain information about processes, sales records, formulas, and the ingredients of their products.

Concerned environmentalists and public-interest advocates distrust the producers' definition of trade secrets and think proprietary rights should never hold precedence over the right of citizens to know what is being used or purchased. They believe that federal regulations on deceptive advertising and sales practices and on warnings against dangerous products do not go far enough to protect citizens.

The basic question is: Why cannot the citizen have information that is collected by the government but is considered a trade secret by the producer? Producers remind governmental officials that the Freedom of Information Act does not apply to trade secrets, industrial processes and operation, style of work or apparatus, confidential statistical data, or the amount or source of income, profits, losses, or expenditures of any person, firm, partnership, corporation, or association. Such disclosures involve civil and criminal penalties for federal employees who make such disclosures.

The law is explicit, but so are the problems faced by citizens who receive massive doses of toxic chemicals. Should a citizen be required to put blind trust in a producer and a government agency? A so-called competitive safeguard is an iron curtain to the consumer—but not to a chemist or, paradoxically, to a business competitor. With a little investment in time, money, and effort, using new, sophisticated equipment, a chemist can identify almost any synthetic material present in a consumer product or emitted into the environment by such a product. This information is available to just about everyone in the business—but not to the poor consumer.

Proprietary rights are not absolute. A landowner can't plant land mines in his fields to discourage trespassers, and a manufacturer can be brought to court over noxious odors emitted from his plant. A consumer must have the right to know what chemical additive is in a hairspray or household cleaner. The consumer's health and safety should not depend on the producer's word alone.

If the citizen is a laborer in a processing plant, doesn't he have the right to know how dangerous the material is that he processes? Shouldn't he have access to his health files? The Oil, Chemical and Atomic Workers have inserted clauses in recent oil company contracts requiring that health files of workers be open to workers themselves. Some dozen major oil companies agreed readily, but Shell Oil refused, and the union, with the backing of a number of public-interest groups, conducted a four-month strike and boycott of Shell products in early 1973, after which the clause was inserted in the union contracts.

The working environment often contains very dangerous chemical compounds. Dye intermediates, such as beta-naphthalamine, are highly dangerous; workers in contact with the chemicals ultimately get bladder cancer. The South Carolina plant manager of the only known producer of beta-naphthalamine was informed in 1972 by the National Institute of Occupational Safety and Health of the extreme danger of his company's product, and the process was immediately replaced by a safer chemical method costing a mere six cents a pound more—a small price to pay compared with near certainty of eventually developing cancer among plant workers.

Consumers and laborers don't have the expertise to judge toxicity, but this doesn't lessen their right to know. Means to analyze and test chemicals are widely known, and several laboratories do *pro bono* work for concerned citizens. However, these willing scientists are often unknown to afflicted parties, and there are limits to their free time. Consequently, there is a great need for stricter governmental regulation of the production, use, and testing of chemicals—as well as a need for an evaluation of how well the government is doing this job.

The growing number of dangerous chemicals makes it imperative that trade secrecy be redefined. In the act of processing laborers have the right to know about toxic chemicals; in purchasing products the consumer has a right to know what is in them; in cases of environmental pollution, those living in the vicinity of a plant should be able to find out what a producer is dumping. These rights of laborers, consumers, and plain citizens should not be abrogated. When a producer declares that a product is safe, he is making a judgment that should be made also by affected citizens. Consumers should refuse to become guinea pigs for producers in determining the toxicity of products—which may become known ten years later, when the consumer is in his grave.

Producers are gradually being forced to honor legitimate requests from citizens concerning toxic chemicals. By federal law no cancer-causing additive is allowed in our food supply. To this extent general welfare and prudence have superseded the tastes of producers of chemicals. Even the presence of ingredients not normally harmful should be made known to the purchaser—as in the case of an ice cream manufacturer who flavors his product with peanut butter, for some people are highly allergic to that otherwise harmless ingredient.

Identifying the chemical name of an additive does not guarantee that the consumer will use the product properly, but consumers should be able to demand warnings, bans, and extensive testing of a suspected material. Mere labeling of trade names and company formulas are not enough (XC-67, for instance, means little to anyone outside the producer's development laboratory). The degree of contact and extent of usage may require that more information be revealed regarding amounts, dispersal media, stability, and solvents. Often the *inert* ingredients in household cleaners and polishes contain effective but highly toxic solvents such as benzene, toluene, and xylenes. Liquid Gold, a furniture polish, may get the wood shiny, but it contains nitrobenzene, which is highly toxic to the skin. In all fairness, the manufacturer of Liquid Gold both lists the ingredients and gives many cautions (in small print) on the can. Quite a few manufacturers don't even do that.

Citizens are entitled to company-generated toxicological data, especially where large quantities of the materials are dispersed into the environment or where people come into direct contact with the chemical. *The more serious the effects of a chemical, the more should be publicly known about it.* The extent of the information supplied should be determined by whether or not the following questions apply:

1. *Do the chemicals cause harm in manufacturing?* A number of asbestos workers have contracted asbestosis due to exposure to this mineral in the manufacturing stage. These workers deserve to know when they are exposed to even small quantities of this material, and what precautions should be taken to guard against inhaling the minute fibers.

2. *Does the process involve emission of hazardous by-products?* A chemical plant on Minamata Bay, Japan, used a mercury catalyst, which was allowed to contaminate the nearby fishing grounds of many poor people living in the vicinity. These unsuspecting fishermen would eat two to three meals of fish per day and thus were severely affected by the mercury contamination. Some died, and others were crippled for life. For several years after the cause of contamination was known, the plant continued to discharge wastes directly into the bay.

3. *Are intentional toxic effects part of the product's commercial value?* Powerful herbicides and pesticides used to kill weeds and pests are produced by the millions of pounds. Some are nerve gases, which are gradually released over long periods of time. Others can hurt and kill unsuspecting citizens who contact or inhale them in a variety of ways. It is important that the dangers

and toxicity of this class of compounds be understood and made known.

4. *Do the chemicals remain in the environment for long periods of time?* Persistence often is a selling point for chemicals, especially those which resist biochemical degradation. Such chemicals as polychlorinated hydrocarbons enter air and water and gradually disperse throughout the ecosystem. Chemicals like DDT and PCB (polychlorinated biphenyls) are commercially beneficial, but, although they have been in use only a few decades, they are found in tissues of animals and fish throughout the world.

5. *Do the chemicals degrade into toxic materials that remain for short periods of time before degrading into other end-products?* Here the problem is the opposite of the preceding. Biodegradable intermediate products are often hard to detect, yet they may be quite toxic to man and animals. Propanil, a major pesticide, degrades into a short-lived intermediate that is highly cancer-causing but proceeds to break down under normal circumstances into harmless final products.

6. *Are the chemical additives used in large quantities?* In a heavily consumption-oriented economy like ours, a new product can be developed and introduced into the market in immense amounts in a short period before adequate testing is completed. A few years ago the detergent industry came under criticism for including both phosphate and caustic ingredients in their products. They announced the use of nitrilotriacetic acid (NTA) in place of these, but an anticipated 2-billion-pound production of NTA alarmed toxicologists, since little experimental information was available about the chemical. When one common food color, Red II, came under attack, the industry converted to Violet I and increased use of Violet I from 3,000 to 66,000 pounds in a single year. In early 1973 the Food and Drug Administration received evidence from Japanese scientists that Violet I material was cancer-causing, and thus it was banned.

7. *Are the chemicals subject to misuse?* Printed regulations and instructions on methods of application, however strict, are insufficient when pesticides and other potent chemicals are used by unsophisticated workers. In early 1972 illiterate Iraqi farmers allowed their families to eat mercury-contaminated seed wheat, and hundreds of deaths and injuries ensued. Parathion sprayed on

tobacco in North Carolina in 1971 killed two farm children. Dozens of deaths have occurred from children sniffing or misusing aerosols containing a dangerous freon propellant. Labels and warnings are hardly enough to keep children from being poisoned by these chemicals now found in many households.

Mere public knowledge of harmful materials will not prevent accidents, but better dissemination of information to the public would make it possible to balance the marginal benefits of such products against the immense risks involved. Removing some items from the market entirely, making safety caps on drug bottles mandatory, requiring licensed applicators in some cases, and banning many aerosols from home use would be the fruits of such information. We are prone to forget that the number of dangerous items within the reach of children is rapidly increasing. Aspirin, a supposedly harmless drug, has the dubious distinction of being the number-one drug killer among children.

8. *Do undetermined amounts of toxic additives accompany known pollutants?* We hear a lot about such air pollutants as nitrogen oxides and carbon monoxide. The gasoline and oil sources of many of these better-known pollutants also contain additives whose identities are held to be trade secrets by manufacturers. Little or no testing of the effects of these billions of pounds of additives has been done. Their investigation has been buried under the sheer volume of research on better-known toxic substances. Tetraethyl lead and other antiknock additives, certain deposit modifiers, and the lead-removal agents (called "scavengers") in gasoline are highly toxic. Some enter the environment directly, while others slip in through careless disposal of waste motor oil and accidental spillage.

Any honest technological assessment of the chemicals used in commercial products depends on the answers to these eight questions. However, we cannot get the answers without breaking down the iron curtain of trade secrecy. A trade secret may be anything useful or advantageous in business that is not generally known or easily and immediately ascertainable to members of the trade. The trade secrecy cloak is a luxury modern man can ill afford in an age of environmental crisis. The Food and Drug Administration requires that ingredients of food additives, drugs, and, more recently, cosmetics be identified on the label. Industries making these products still thrive and make handsome profits. Safeguards for protecting proprietary rights already exist within our patent laws. One needn't expect dire consequences to befall the fuel, tobacco, liquor, and detergent industries if the law were to require that ingredients be revealed.

As noted earlier, sophisticated instruments and techniques have been developed in the last decade or so that greatly simplify product analysis and detection: mass and nuclear magnetic resonance spectrometry, scanning electron microscopy, gas chromatography, and many others. Certain natural biochemical products defy easy analysis because of their complexity, but with enough money and time they could be characterized also. It is the public's limited resources that fail to breach the walls of secrecy, not the competitor's. Producers who mask products with extraneous materials do not fool chemical detectives. However, most scientists would agree that reverse engineering—working backward from product to process—is extremely difficult. When dealing with complex mixtures like perfumes and liquors it is nearly impossible, since mere traces of exotic compounds provide that extra flavor or aroma. Quite often the perfume or liquor maker is not aware of the composition of all the ingredients, and so such knowledge is academic.

The attack on trade secrecy is also an attack on the consumption culture. The public does not want to be a guinea pig of profit-making corporations. If an individual curbs his own appetite, he is still affected in some way by what his neighbor buys. Who has a final say on benefit and risk? Producers, consumers, or the general public? A growing environmental consciousness will not tolerate the naive notion that a product is safe until proven harmful.

Quite often a product can't always be proven conclusively safe. However, prudential decisions and proper technological assessment must be made, and these require value judgments. One such judgment is that the public may not tolerate excessive use of luxuries when these threaten health and well-being. This is a social value judgment that needs further elaboration. Basic principles governing such judgments in regard to consumer products should include:

1. The consumer's right to know what he has purchased, in what quantity, and whether it has any harmful contaminants.

2. The consumer's right to refrain from buying what he deems is harmful to himself (freedom to choose or refuse).

3. The consumer's right to complete labeling of products and to access to performance standards and complaints, if any, from other users of the product.

Basic principles for proper value judgments in regard to environmental contaminants should include:

1. The public's right to know what is being emitted into the environment, what quantities are present, and how toxic the pollutant is.

2. The workers' right to know what chemicals they are working with and the right to periodic medical examinations, with access to their personal files.

3. The public's right to require pretesting of all new chemicals, especially those belonging to families of compounds known to be toxic.

4. The public's right to regulate production of any material that is harmful or can produce harmful by-products.

The fulfillment of all of these demands will cost the producers money. Production of some products will be delayed during testing, and some will be eliminated due to suspected toxic effects. This is a small price to pay compared to the possible cost in human health. Today, products about which little is really known (except their ability to yield a profit) flood our markets by the thousands. Safer and better products will become available only if the public is better informed, and that can only take place when the iron curtain of trade secrets has been successfully breached.

The Need for Technology Assessment*

A free choice of alternatives and a free flow of information are essential before citizens can make rational technology assessments. On a number of occasions in recent years, technology assessments,

---

* Technology assessment is an attempt to evaluate in advance the environmental results of applied technology, in order to develop policies for its rational use.

if properly made, would have proved most useful. In the early 1970s the United States decided to discontinue development of the supersonic transport after considerable money had been spent on research and development. While environmentalists were pleased with the final decision, the process of technical assessment left much to be desired. Information was suppressed and withheld from inquiring citizens. Special interests exerted strong pressures on members of Congress. Health questions and noise issues were never clearly expressed, and it is more than likely that the most influential factor in the decision to discontinue development was the fact that the airlines were having second thoughts about these large, expensive, unproved aircraft. Congress's decision was based on a most rudimentary technology assessment.

More complete technology assessments must be applied to other problem areas besides aerospace. Judgments are urgently needed in such instances as energy consumption in aluminum-making, communication satellites, release of computer information, solar power generation, and nuclear power technology, to name but a few.

An orderly technology assessment must include setting priorities on technological problems, establishing an effective assessment procedure, and providing for utilization of assessment results once they have been completed. Priorities should not be made on the basis of political or economic expedience. One industry should not be given preferential treatment over another simply because a powerful congressman wants it located in his state. The size of the investment or the political power of an industry should not be prime factors in making an honest assessment—neither should the need of jobs for members of a powerful labor union. Technological problems that affect the largest number of citizens and have the greatest environmental impact or require the consumption of excessive amounts of natural resources should always be given top priority.

Congress's Office of Technology Assessment (OTA) is finally beginning to function. One of its first acts should be to establish a priorities board, composed of experts from various scientific disciplines and of private citizens having no interests in the projects or industries to be assessed. Board members should request and examine all pertinent records already available, such as reports issued by the National Academy of Sciences and the National Academy of Engineering, environmental impact studies, and preliminary studies made by the concerned industries. When no environmental impact statement, as required by the National Environmental Protection Act, is available, the proponents of the new technology should be required to furnish the requisite information.

Often it cannot be ascertained who will benefit the most from the new technology. In such a case the general trade association involved—say the American Petroleum Institute—may be commissioned to write the preliminary report in the name of its member companies. Initial funding may come from the OTA and eventual reimbursement from the industries that ultimately use the approved technology. Public initial funding will not be ill-spent in the light of the many costly errors that have been made in recent years due to lack of assessment.

Once the priority list has been established, the OTA should set in motion an adversary process, in much the same manner in which legal matters are argued before the courts to present both positive and negative aspects of the new technology. First, all the positive arguments should be assembled, including economic benefits, numbers of prospective employees, improvements over existing methods, ways in which the innovation will improve quality of life, how widely the benefits will be felt by the nation, and whether these can be extended to other peoples in the world. Negative arguments should be mustered next. For this purpose public-interest groups opposing the project may need funds to travel to the hearing site, and these should be provided by the Office of Technology Assessment.

After the OTA board of judges has determined that all reasonable means have been used to elicit arguments for both sides of the question, the two positions should be open for cross-examination. Should the opposition to the innovation not have sufficient manpower, devil's advocates should be commissioned by the OTA. If such opposition has considerable resources, coming, say, from a competing or threatened industry using traditional technology, a citizen's advocate should be appointed. The entire cross-examination should be public and should allow for further comments from the general public before the formal examination time is terminated.

The OTA judges should draft a report from the information thus obtained. It may be obvious from the report that certain safeguards or modifications are needed before the project can be approved, or that it is a major hazard to man and his environment and should be canceled. Or it may be that the judges can come to no judgment and must request a second set of public hearings and repeat the adversary process. Opposition arguments may require elaboration, and further substantiation of claims may be required from the producer industry, in which case the second adversary procedure may have to be postponed until new research data is available.

The philosophy behind adversary procedures may seem strange to persons engaged in "objective" scientific work. However, this is much the same process of critical review that is done behind the scenes with research papers in journals, except that the procedure is public. The naive notion that a panel of objective technical experts will render an unbiased opinion belies the fact that citizen acceptability of possible risks must be a determining factor. Adversary procedure holds that there is no simple "yes" or "no" to any new technological process. Benefits clash with risks, and both must be given public airing. Debate sharpens the focus, reveals weaknesses and problems, and suggests remedies previously unseen or not admitted. The adversary method has a democratic flavor, for it ensures that judgments are not made by objective technologists—who often have unconscious biases—but emerge in public confrontation.

Draft copies of the report on the assessment should be circulated to all interested parties and made available for public inspection. After weighing all comments, a final report should be prepared and published. It is quite likely that a minority report will also have to be prepared as an appendix to the final report. The final report should include all recommendations for both new legislation and further regulation under existing governmental agencies.

Through a rational and public assessment of technological innovations, citizens will have a voice in how future production and consumption will proceed. But assessment should not just be limited to national policies. Global environmental impact statements should be required for critical review by some United Nations organization so that a global technology assessment can be effected.

National and international institutional structures for assessment are not enough. Citizens must be willing to propose, discuss, weigh, and determine strategies and substrategies on group and individual levels. The application of U.S. intellectual resources to creating and perfecting these strategies represents a qualitative leap from the days of research and development geared exclusively to greater profits and increased consumption. It marks the beginning of a global conservation policy.

5 · Conservation Strategies

Conservation begins at home but doesn't end there. To conserve our natural resources we need a policy that is comprehensive, covers all segments of the population, and includes all major areas of energy and materials expenditure. We can't expect a few valiant citizens to bear the whole burden of conservation alone. A healthy democratic society needs citizens' involvement in government. So does a healthy physical and social environment.

Americans prefer encouragement by incentive to restriction by regulation. However, both carrots and sticks are effective. Thus, the following four sets of strategies should increase the citizen's awareness of the environment, encourage him to cooperate on local and regional levels, help him to apply pressure at the federal level, and open his eyes to the need for international cooperative ventures.

Individual Strategies

Conservation practices are good not only for the other guy, the other company, the other nation. They are good for each of us as individuals. And they can save us money besides.

Our individual actions in the home are indications of our real attitudes. We may turn off a few more lights tonight than we did before the energy shortage or refrain from using green toilet paper, but that's not enough. Let's look at various day-to-day practices that consume energy and materials and identify the major areas of consumption.*

---

* For a more precise estimate of how much energy and materials we consume annually as individuals, turn to the Lifestyle Index at the end of this book.

Residential Energy Conservation

It should come as no shock to the reader that Americans use enormous amounts of electricity in their homes. Americans are quick to elevate luxuries to the rank of necessities. This practice has been vividly pointed out by Wayne King in an article in the *New York Times* (April 22, 1973), about the household of Bob and Alice Alotta of South Philadelphia. While this family of four considers themselves working-class people, they use an impressive number of electric appliances in their home: electric blender, coffee maker, freezer, electric frying pan, two irons, eight radios, a refrigerator, two black-and-white television sets, two toasters, three vacuum cleaners, two small record players, a more elaborate stereo set, two electric toothbrushes, two electric clocks, two tape recorders, an electric sewing machine, an electric "water pik," an old electric train, and a wood-burning set.

With the proper data the reader can make his own personal survey of what is needed to operate his home (the key word is *needed*). Perhaps your inventory will reveal more items than the Alottas'. However, the number of appliances is less important than the length of time they are used, and a budget-conscious family may constantly harp about excessive use of electric lights, while neglecting other conservation practices that could generate greater energy savings.

Potential household savings can be divided into three classes: heat savings, cooling savings, and electric appliance savings. Good heating conservation practices can save up to half the energy bill in states where winters are moderate to severe, and proper insulation can help cool homes where summers are long and hot.

The Office of Consumer Affairs, the American Gas Association, and Concern Incorporated of Washington, D.C., have compiled hints for fuel reductions in both heating and cooling.

HEATING

1. Weatherstrip and caulk around windows and doors. Between 15 and 30 per cent of all heating bills can be traced to warm air leakage.
2. Install storm windows or insulating glass. This will pay for itself in seven years—an annual return of 18 per cent on the investment.

3. Install overhead and side-wall insulation. This type of insulation will pay for itself in one or two years.

4. Clean heat-exchange surfaces in your heating plant for greater efficiency.

5. Seal against air leakage in attic and from unoccupied spaces.

6. Close the damper in unused fireplaces.

7. Close window drapes in cold weather and, especially, in cool windy weather. Also close draperies at night when the temperature is 60 degrees or less.

8. Close doors in unused areas. Avoid continual opening and closing of doors in used areas.

9. Refrain from putting objects in front of heat registers.

10. Lower thermostat settings at night. Fuel savings can amount to three-fourths of 1 per cent for each degree reduced. Teach children to sleep in cooler rooms in winter and emphasize that warm covers are as good as a warm room.

11. Never turn the thermostat higher than seventy-two degrees. It is not good for one's health, and one can always adjust to individual needs with a warm sweater.

Insulation practices that apply to heating the residential unit also apply to cooling. Good building practices result in a cool house. The deep eave extensions used in tropical and pre-air-conditioned homes furnish shade and a cooling effect absent from many modern homes. If your house lacks eaves, there are still some significant savings to be achieved if the following suggestions are carried out.

COOLING

1. Avoid air conditioning if you can. Many people complain of summer colds that seem never to go away. Ask them if they have tried turning their air conditioners off.

2. If you must have air conditioning, install units in occupied spaces only and avoid use when the weather is tolerable.

3. Make sure that all windows open, so that you may benefit from seasonally cooler weather.

4. Buy or rent a building with less glass.

5. Paint the roof of your home a light color. This reflects the sun's rays.

6. Clean air-conditioning filters as necessary and replace regularly, or at least seasonally in temperate climates.

7. Keep windows that face the south and west shaded in summer to prevent solar heating.

8. Don't let lamps and appliances that generate heat be near a thermostat. This causes more cooling than needed.

9. Don't place objects near cooling registers, causing the system to work harder and longer than necessary.

10. Close off areas not used in summer.

11. Don't turn the thermostat too low. An ideal temperature is seventy-two degrees (seventy-eight degrees was recommended for the 1974 energy crisis).

12. Use dehumidifiers when possible to save on energy. (Often discomfort is due to excess humidity rather than heat. Remember that house plants serve as good humidifiers.)

13. When leaving on summer vacation, be sure to turn off the air-conditioning unit. This applies to absences during the day or weekend as well.

Conservation of electric power in the home requires that we ask ourselves several questions about electric appliances: Do we really need this or that appliance, or is our purchase due to advertising and social pressures? If we must buy it, which type should we buy? Once installed in the home, how much use should we make of it?

A sizable portion of American homes have many of the twenty-four major classes of electric appliances listed in Table VII. Once one decides to purchase an appliance, a good conservation practice is to compare power and energy demands of the various models offered for sale. This is not easy, since many brands do not post energy requirements. (Average energy requirements for appliances are listed in Table VIII.)

Consumers don't realize that many appliances are really energy hogs. For instance, a frost-free refrigerator requires 50 per cent more energy to operate than a standard model and costs $3–$6 a month to operate (compared to $2–$4). Side-by-side refrigerator-freezers use about 45 per cent more energy than the conventional models. Some early dishwashers had drying cycles that could be manually cut off when desired. Dishes air-dry with immense energy savings. Modern apartment models of dishwashers, built

TABLE VII
Electric Appliances in U.S. Homes, 1972

	Homes Wired (*In per cent*)
Room air conditioners	46.7
Bed coverings (electric)	52.7
Blenders	41.5
Can openers	49.5
Coffee makers	93.1
Dishwashers	32.0
Dryers	51.0
Disposers (food waste)	31.9
Freezers	34.3
Frying pans	59.4
Hotplates and buffet ranges	25.0
Irons (total)	99.9
Irons (steam and steam/spray)	91.4
Mixers	85.2
Radios	99.9
Ranges (free-standing)	44.1
Ranges (built-in)	16.6
Refrigerators	99.9
Television (black and white)	99.8
Television (color)	60.7
Toasters	95.1
Vacuum cleaners	96.9
Washers	96.9
Water heaters	36.3

SOURCE: *Merchandise Week,* cited in *Statistical Abstract of the United States, 1973,* p. 693.

TABLE VIII
Estimated Power Consumed by Electric Home Appliances in a Year

	Average Wattage	Kilowatt Hours Consumed Annually (*Estimated*)
FOOD PREPARATION		
Blender	386	15
Broiler	1,436	100
Carving knife	92	8
Coffee maker	894	106
Deep fryer	1,448	83
Dishwasher	1,201	363
Egg cooker	516	14
Frying pan	1,196	186
Hot plate	1,257	90
Mixer	127	13
Oven (microwave)	1,450	190

Conservation Strategies

	Average Wattage	Kilowatt Hours Consumed Annually (*Estimated*)
Range (with self-cleaning oven)	12,200	1,207
Range (with regular oven)	12,200	1,175
Roaster	1,333	205
Sandwich Grill	1,161	33
Toaster	1,146	39
Trash compactor	400	50
Waffle iron	1,116	22
Waste disposer	445	30
FOOD PRESERVATION		
Freezer (15 cubic feet)	341	1,195
Frostless	440	1,761
Refrigerator (12 cubic feet)	241	728
Frostless	321	1,217
Refrigerator/freezer (14 cubic feet)	326	1,137
Frostless	615	1,829
LAUNDRY		
Clothes dryer	4,856	993
Iron (hand)	1,008	144
Washing machine		
Automatic	512	103
Nonautomatic	286	76
Water heater		
Standard	2,475	4,219
Quick recovery	4,474	4,811
COMFORT CONDITIONING		
Air cleaner	50	216
Air conditioner (room)	860	860
Bed covering (electric)	177	147
Dehumidifier	257	377
Fan		
Attic	370	291
Circulating	88	43
Rollaway	171	138
Window	200	170
Heater (portable)	1,322	176
Heating pad	65	10
Humidifier	177	163
HEALTH AND BEAUTY		
Germicidal lamp	20	141
Hair dryer	381	14
Heat lamp (infrared)	250	13
Shaver	14	1.8
Sun lamp	279	16
Toothbrush	7	0.5
Vibrator	40	2

TABLE VIII (continued)

	Average Wattage	Kilowatt Hours Consumed Annually (*Estimated*)
HOME ENTERTAINMENT		
Radio	71	86
Radio/record player	109	109
HOUSEWARES		
Clock	2	17
Floor polisher	305	15
Sewing machine	75	11
Vacuum cleaner	630	46

SOURCE: The Electric Energy Association.

into the sink and cabinet systems, generally have automatic drying cycles and consume five times the electricity of the old-fashioned models.

The following practices will cut down consumption of energy by electric appliances in use around the home.

ELECTRIC APPLIANCES

1. Get rid of unnecessary appliances, whose mere presence is a temptation to use electricity. Disposal units are environmentally wasteful, and compressors for garbage can easily be replaced with a small expenditure of human energy. Homes have too many electric clocks, and power failures make these less dependable than the mechanical types.

2. Slight changes in household practices can make some electric appliances superfluous. Charcoal lighters, ornamental lamps, electric toothbrushes, hair dryers, and electric can openers and combs fall into this category, as do meat carvers and electric saws, which are dangerous besides.

3. Avoid excessive use of washers, dryers, and electric irons. Purchasing duplicate clothing items and operating washers at full capacity only could save a third of such washings per year. Use washers and irons before peak loads in your community (before 8 A.M. and after 6 P.M. as a rule).

4. Run the clothes washer on cold water when possible. Hot water faucets should be repaired immediately when leaky. A simple steady drip from a leaking faucet can waste up to 700 gal-

lons of water a year. Never wash dishes under running water. Teach children not to run water while brushing teeth or washing.

5. Refrigeration units should be left closed as much as possible. Defrost refrigerators regularly, and check door gaskets for tight seal. If you use a freezer, remember a well-stocked freezer requires less energy to operate than a partly filled one.

6. If you must have an electric stove—gas stoves are cheaper—avoid models with self-cleaning ovens (each cleaning operation uses about sixty-five cents worth of electricity at 1973 rates). Match the size of pans and pots to the burners to avoid wasted heat, and defrost foods before cooking.

7. Turn off television and radios when not in use. Buy a solid state TV when replacing the old set and try to get along with black and white. If you use a tuning device, unplug it when not in use; it consumes electricity all day long.

8. Install fluorescent lighting; it is more efficient than incandescent lighting. The former converts about 20 per cent of the electrical energy to visible light, the latter about 5 per cent. Turn off incandescent lights when not in use for short lengths of time (this does not apply to fluorescent lights).*

Gasoline Savings

The chances are better than one out of two that the reader owns a car and has watched the gasoline bills go higher and higher. Gasoline, which averaged less than forty cents a gallon in 1970, will perhaps be a dollar a gallon before the end of the decade. With an average motorist buying more than 700 gallons per year, gasoline looms as a larger budget item than ever.

In the past producers of the major brands have made little effort to teach the consumer how to trim his or her gasoline budget. A few companies are making feeble stabs at conservation messages in their advertising, but it is due more to public relations than conviction. The lack of good conservation information on gasoline use, except for a brief spell during the Arab oil embargo,

---

* *A note of caution.* Home economies can be penny wise and pound foolish. Energy economies should not be made at the cost of eyesight. We cannot expect youngsters to read in Abe Lincoln fashion before open hearths. Savings should not be made at the expense of nerves either. Too many petty savings might become an excuse for badgering members of the household. In such a case, a little more appliance use might be worth the extra peace and quiet.

is responsible for our runaway consumption of this fuel in the late 1960s and early 1970s. Some practical hints are in order:

1. *Keep the car at home as much as possible.* Granted one needs the car for intermediate and longer trips, what about local trips to buy milk or bread? Half of all auto trips are less than five miles. A recent study for the National Science Foundation by Eric Hirst and Robert Herendeen shows that these short trips account for 11 per cent of total auto mileage and about 20 per cent of urban mileage. The energy expended on half of these short trips could easily be replaced by muscle power.

Walking shoes or a trusty bike are the best gasoline-saving equipment you can buy. Human muscle can save up to 5 per cent of your gas bill. An added enticement is that walking and biking are good exercise and require less than 4 per cent of the energy needed to drive a car. Besides, most auto dents and scrapes occur on short trips. Adopt the motto: Learn not to have to drive.

2. *Drive conservatively at all times.* When two drivers use the same car, there may be as much as 10 per cent difference in gasoline mileage. Tender loving care can be economical. Hard driving leads to poor fuel economy. Rapid acceleration, jumpy driving, racing an idle motor, high cruising speeds, fast starts, and a preference for heavy intracity traffic all take a toll on your gasoline bill. An even cruising speed of around fifty miles per hour is best for fuel economy. The Department of Transportation suggests: Avoid pressing the accelerator all the way down when climbing hills and long grades; turn off the motor if you stop for more than a minute; look ahead and pace yourself to minimize stops at traffic lights and in jam-ups.

Safe (and thrifty) motorists keep their cars in top condition. Regular servicing with tune-ups as recommended (usually between 5,000 and 12,000 miles) will result in major savings. Your serviceman should be asked about faulty valves, piston rings, and carburetors, for these cost fuel. Tires should be checked regularly for proper alignment and air pressure. The rolling resistance of tires affect gas mileage adversely. The extra cost of radial steel tires will favor your pocketbook by resulting in a noticeable saving in fuel.

3. *Review your gasoline-buying habits.* Brand loyalty often outweighs economy. Reasons for this loyalty may range from

habit, friendly service, clean rest rooms, and the convenience of credit cards to an unconscious attraction to the color or design of the company's signs. Traditionally, gifts and advertising come-ons have influenced customers more than gasoline quality.

There is no tiger in your tank. You are buying the same gas from Cheapo as from the higher and better-known brands down the street. Fuel pipeline operators admit they interchange various brands of gasoline. Strict company control of fuel from well to station pump is the exception rather than the rule. A tankful of gas will vary more by season and region than by brand.

Buy gasoline by price. The gas wars of yesteryear are gone. The great surpluses of fuel sold in bulk to independent dealers is a thing of the past. Still, some filling station operators manage to survive while selling at lower prices because they do not have fringe services. Be on the lookout for these.

Buy the lowest octane fuel that will keep your car from knocking. Experts estimate that 40 per cent of all drivers purchase fuel of too high an octane rating. Don't believe the myth that premium means better, or that an occasional tankful of high test will blow out the carbon. Owner's manuals recommend specific octane grades, but these refer to the average car coming off the assembly line. Yours may be above or below the average. Furthermore, your car's octane requirements may change with age and engine build-up of lead.

A thrifty auto owner knows how to experiment. When his car begins to knock, especially on steep climbs, it is a sign that a higher-octane grade is needed. If it doesn't knock, the octane may be too high. Try the next lower grade and continue until you detect knocking. Most 1971 and later cars, and some earlier ones, use low-lead ninety-one octane (RON) fuel. Don't hesitate to try mixtures of two grades, say half and half, or a quarter of higher octane and three-quarters of lower. Choose a responsible dealer who has a heavy gasoline turnover (not a difficult feat these days). Fuel that sits in the station can accumulate dirt, water, and harmful polymers, which affect auto performance. Ask if the station's pumps have filters. Find out how dependable your service station people are. Prompt and courteous service says much about fuel tank maintenance.

4. *Buy a small car known to have good fuel economy.* Oil companies admit that exotic additives are not the major deter-

minant of gasoline efficiency—but the auto itself is. The Environmental Protection Agency's report *Fuel Economy and Emission Control* says that the vehicle's weight is a major factor. Decreased fuel economy is partly due to making cars of the same model name heavier over a period of years. For instance, in 1958 a Chevrolet Impala weighed 4,000 pounds and got 12.1 miles per gallon; a 1973 Impala weighs 5,500 pounds and gets only 8.5 miles to the gallon.

Weight and engine size are not the only factors in fuel economy. Accessories and super-duper features like push-button windows also take their toll. Air conditioning can amount to an average 9 per cent of your fuel bill and can go to 20 per cent, depending on weather and traffic conditions. Automatic transmission can account for 7 per cent or as much as a good emission control device. Power brakes and power steering cost fuel; so do design changes in exhaust system configuration, carburetor design, and compression ratios. Some alternative engine designs, like the current rotary motor, use more fuel than standard piston engines.

Watch out for gasoline guzzlers when buying a car. It could mean a difference of a thousand dollars during the life of the car. Ask motorists who drive the car you are considering buying how many miles they get per gallon. And don't confuse miles per gallon with miles per tank. One major auto-maker cut complaints in 1972 by simply increasing the size of the gas tank, after it was discovered that the disgruntled customers had been counting service station stops rather than adding up their gas bills.

Materials Conservation

The ultimate user of materials is the individual citizen or some social unit working for the individual. While industry uses a third of our energy and processes most of our raw materials, it is the individual consumer who drives the auto and turns on the electric appliances. Consumption patterns have been developed at great expense of time and money and are not easily changed. It is generally not profitable for industry to encourage re-use of materials. Televison commercials show housewives pouring out heaping measures of detergents, although instructions on the box say to use far less.

American military personnel in Japan and visitors to that country after World War II were amazed to see every metal can grabbed up and converted into some useful object or into a salable trinket. Japanese culture teaches frugality. Many Americans who experienced the Great Depression of the 1930s recall how their families conserved on food, clothes, and other items. Some had their electricity disconnected, others their telephones. No one hopes that that unfortunate era will be repeated, but some of its conservation practices are worth investigation.

Cynics say there is no point to Boy and Girl Scouts scouring the countryside for discarded cans and tires or turning civic centers into junk piles. Recycling programs, they say, are successful only to the extent that the junk can be re-used. American youth can learn conservation through the time and effort taken to collect junk, but such education programs have little value when the individual is being taught at the same time in hundreds of subtle ways to consume more. Material consumption grows in this country at the rate of 4 to 5 per cent per year, while the rate of population increase is less than 1 per cent. The per capita consumption of materials in the United States was 22 tons in 1965, 24.7 in 1968, and 28 tons in 1971. In addition, we are in danger of being buried under mountains of packaging materials. Everyone is familiar with the aftermath of gift openings at birthday parties and at holiday time, when someone eventually has to go around and gather up the pretty ribbons and colored wrappings that have been discarded. This same purposeless waste is repeated each time the salesclerk unnecessarily wraps a purchase. According to a recent Public Health Service publication, per capita packaging consumption has steadily increased in the United States from 404 pounds in 1958 to 577 in 1970.

Production of packaging materials and containers is a lucrative business. Plastics, aluminum, paper, and especially glass are used by the ton,* often for purely cosmetic purposes. Both plastics and

* A 1973 Environmental Protection Agency *Report to Congress on Resource Recovery* claims that packaging and containers were major users of the following materials in the years indicated:

	Tons Used in Packaging	Per Cent of Total
Plastics (1969)	1,729,000	20
Glass (1967)	8,900,000	70
Aluminum (1969)	183,000	14
Paper (1967)	1,400,000	2.5

aluminum are environmentally costly. Plastics are hard to recover, require nonrenewable petroleum products, and often contain toxic plasticizers. They defy degradation by natural processes.

Aluminum and bimetallic cans require enormous amounts of energy to manufacture; the costs are definitely not internalized. The user is free to discard cans whenever and wherever he pleases; the cans stay around for years, and flip-tops are dangerous to bare feet. However, aluminum, unlike plastic, is quite valuable as scrap. It is worth at least $200 a ton and is in great demand, but, unless we add a high deposit to the price of the product in the can, cans will continue to be discarded in large quantities. Outlawing the use of certain materials in containers may be the ultimate conservation technique.

Citizens can reduce material consumption in the following ways:

1. Refuse unnecessary packaging in paper bags and wrapping paper at stores.

2. Use aluminum foil sparingly and re-use where possible.

3. Choose products with the least packaging and buy items in bottles and jars that can be reused. Don't purchase foodstuffs in containers with fancy shapes that must be thrown away with 10 per cent of the contents still inside.

4. Buy drinks in returnable bottles and containers only.

5. Avoid TV dinners and snack foods.

6. Buy from produce centers and vegetable dealers who use a minimum of packaging.

7. Prod food manufacturers to refrain from packaging with aluminum. Organize boycotts when such materials are used wastefully.

8. Organize sales of used furniture, kitchen utensils, clothing, and assorted junk and recycle household items into another home. People are always looking for bargains.

9. Don't buy "disposable" paper products such as facial tissues, paper napkins, diapers, plates, and towels. The paper industry is one of the three largest consumers of energy, and paper products require much energy to make (see the Lifestyle Index).

10. Never use aerosol sprays in the home or for personal use. Besides being a very heavy user of energy the aerosol can is unsafe and may prove harmful to children.

To conserve materials is to conserve energy. A University of

Illinois study by Bruce M. Hannon shows that returnable bottle systems use only about 28 per cent of the energy of the throwaway bottle or the bimetallic can. In the manufacture of paper, it takes far more energy to cut trees, haul them to pulp mills, and convert them into paper than to use paper as a starting material. According to a 1972 study of the Midwest Research Institute, it takes 60 per cent less energy to manufacture kraft pulp from de-inked and bleached wastepaper than from virgin pulp. Convenience-food stands seldom recycle their containers, yet a recent study showed that it takes 315 square miles of forest each year just to keep one of the large convenience-food chains supplied with paper packaging.

It may come as a shock to those who think recycling is a recent vogue to learn that proportionately less material is recycled today than two decades ago. Paper waste reclamation declined from 27.4 per cent in 1950 to 17.8 per cent in 1969. Textile reclamation has fallen off because of the introduction of synthetic padding materials. Rubber reclamation declined not only relatively but absolutely, accounting for only 9 per cent of production in 1970. General Motors, a major user of rubber, decided that tires used on its 1975 model year production would contain no recycled rubber. Motor oil re-refiners are in a depression due to high costs both of reclaiming detergent oils and of antipollution equipment.

Collective efforts are needed to control the flood of discarded junk, garbage, and general wastes. Individual strategies and practices are limited by our consumption culture. A massive return of waste materials to the production cycle will meet with determined resistance from primary producers. Furthermore, it is inconvenient to sort and gather materials and transport them to recycling centers, which are difficult to reach in many cases.

Home Gardening

Many people have green thumbs but have only become aware of it in recent years. During World War II 20 million "victory gardens" produced about 40 per cent of the vegetables grown in this country. The early 1970s saw a renewed interest in home gardens, and the number increased to about 24 million in 1972, 27 million in 1973, and over 30 million in 1974, according to the National Garden Bureau of Gardenville, Pennsylvania.

Raising homegrown fruits and vegetables can be fun as well as excellent exercise and good conservation. Cultivation by hand results in a considerable saving in energy (see the Lifestyle Index for actual savings). If you prefer produce that is pesticide-free and perhaps free from chemical fertilizers, then homegrowing is a sure guarantee of getting what you want. A return to the soil not only brings the city gardener closer to nature but broadens his interest in ecological processes and programs and makes him aware of the need to treat nature gently.

Home gardening is good economy. Three dollars worth of seed can yield $100 worth of vegetables. On a plot as small as ten by fifteen feet, vegetables can be so spaced and arranged (small vegetables to the south and west, bushes and vines behind) as to yield one-third of the yearly demand of two people. Gardeners susceptible to modern trends in folklore might try talking to the plants when they appear in the hope that tender loving care will make them grow a little faster. If time, space, and energy allow, home gardening should be made a family affair.*

Community Strategies

Individual and familial savings on materials and energy are simply not sufficient in themselves for a viable conservation policy. They must be coupled with local and regional cooperative ventures, whose encouragement and coordination are beyond individual resources. There are several levels of community action. One is a joint educational and motivational program to alert individuals and assist those who want to know how or where they can conserve. A second is citizen pressure to prohibit wasteful practices or to ensure that collected materials are truly recycled. A third level is wholehearted support of local or regional regulations that improve the environment.

* On the Kentucky farm where I grew up in the 1930s and 1940s, our family grew almost all the food we ate. We grew or gathered: apples (four varieties), plums (green, red, and damson), pears, peaches, cherries, strawberries, blackberries, grapes, rhubarb, head and leaf lettuce, kale, mustard, endive, sweet and white potatoes, tomatoes (four varieties), cucumbers and pickles, carrots, parsley, horse-radish, red and white radishes, onions and chives, beans (green, lima, and kidney), peas, eggplant, sweet corn, turnips, cabbage, spinach, dandelions, and pumpkins. The farm diet also included milk, cheese, butter, eggs, chicken, rabbit, pork, beef, walnuts, and hickory nuts.

Citizens can apply pressure to discourage overheating or overcooling of public buildings, buses, theaters, restaurants, and supermarkets. They can work to close display and refrigerator units in supermarkets, to do away with display lighting, and to see that local ordinances are passed against such energy-wasteful luxuries as electrically heated sidewalks.

Collective citizen action can prod municipal governments into regulating the use of autos in inner-city areas by banning on-street parking, levying a municipal parking tax, or forbidding single-passenger cars in certain areas. Governmental regulations are sometimes disliked, but very often they are necessary to preserve the environment. A good example of this occurred when a conservation group in northern Virginia opened a large tract of land for the enjoyment of ecology-minded groups. In the spring of 1973 a local high school graduating class discovered the tract, made an end run around a barrier, and blazed a trail up to the scenic sites on Bull Run Mountain. Within three months cans and other refuse appeared everywhere. In such a situation appropriate regulations and policing are necessary to save the environment.

Urban Transit Systems, Biking, and Pedestrianism

Two American movements—the fight for individual rights and the struggle for social responsibility—collide head-on when it comes to municipal transportation. In 1973 the Environmental Protection Agency required each state and local government to submit plans for implementing reduction of air pollution caused by cars in congested areas.

The various strategies considered included: traffic controls (express bus lanes, graduated tolls depending on number of passengers, free minibuses; parking controls (taxes, bans, fees); mandatory testing and inspection to expose excessive auto polluters; retrofit systems for older vehicles and antipollution devices on newer ones; staggered work hours to reduce rush-hour taffic jams; and conversion of taxis and government fleets to less-polluting liquified petroleum gas or compressed natural gas.

Whereas automobiles require 8,100 British thermal units per passenger mile in urban driving, walking takes a mere 300 and biking 200. American tourists in Amsterdam are constantly amazed at the almost overwhelming use of bicycles in that beautiful and quaint, though densely populated, city laced with canals. Holland has only one car per five people, compared to one per two people in the United States. Nevertheless the downtown streets of Amsterdam would qualify as a disaster area if the Dutch adopted American transit habits. On a summer morning one can count about fifty bikes on any given city block, moving noiselessly and faster than most cars in urban traffic. American commuter hypertension is lacking, the riders are cheerful, and the air is clean.

The lack of bike lanes, racks, and parking protection in the United States discourage many people who would prefer to ride two-wheeled vehicles to work. On an average, weather permits bike-riding throughout 80 per cent of the year in this country, but present policies favor the auto driver, and few privileges and conveniences are offered to the frustrated biker, a growing number of whom are challenging this policy.

Let's face it, our economy brands the bike as sporting equipment at best and as a toy at worst because it takes so little to produce and operate. Both automobiles and bicycles last about the same length of time, barring unforeseen mishaps, but car maintenance will cost $1,000 per year and maintenance for the bike less than $10—and initial cost of most cars is at least $3,000, whereas a good used bicycle can be purchased for less than $100. The difference in cost is not peanuts, especially in a consumption-oriented economy.

It would be hard on certain businesses if we were to convert to a bicycle economy, for this would mean a 98 per cent reduction in travel spending. Beware, American Roadbuilders Association! No more new superhighways, though you might have to build more bike lanes and underpasses. There would be fewer filling stations and repair shops, but there would be cleaner air, fewer traffic jams, and less nervous strain. Covered walks, less congestion, and auto-free streets could revive our rotting downtowns. In 1900 a horse and buggy could cross Manhattan in less time than a car can today. It's time-saving to walk rather than ride in many of our urban areas. The bike age might usher in an age of the pedestrian as well.

A growing number of Americans would like to see rapid, clean, cheap mass transit; shopping districts free of autos; and more areas where it would be safe to ride bikes and walk. In 1972

Oxford Street in London was closed off to downtown traffic, except for buses and taxis. It had once been one of the most dangerous and polluted streets in Europe. Accidents dropped from forty to twenty monthly, and carbon monoxide levels dropped from twenty-two to three parts per million. However, traffic clogged the neighboring streets, and accident rates increased there. Moral: Localized bans are not effective without citywide and regional policies to back them up.

Community Waste Utilization

A host of junkyards and roadside dumps, not to mention general street litter, remind us of our local solid-waste problems. What are we to do with all the litter we collect? Good landfill sites are disappearing and garbage-collection costs mount. Some communities heap the garbage and cover it with dirt. Some of these heaps are so high that they furnish hill-less areas, like the environs of Chicago, with local ski slopes. The meadows of New Jersey have been buried under garbage, and San Francisco fears its bay will soon be filled, if dumping of solid waste continues unabated.

In 1974 solid-waste disposal costs ranged from $5 to $50 a ton. Productive research is lacking in this area, yet in 1973 the Environmental Protection Agency's Solid Waste Management Program was almost eliminated by the Administration's budget-watchers. New disposal methods being evaluated include incineration plus electric generation, residue recovery, steam recovery, pyrolysis, sanitary landfill, composting, and fuel and materials recovery. Costs vary with the size of the operation.

Incineration by itself is a source of pollution unless expensive devices are installed to trap polluting agents. Materials recovery is good economics, but many communities are too small to afford the costly machinery needed to sort waste materials automatically. Regional cooperative programs are called for unless such communities initiate manual sorting of waste materials at the point of origin. About 43 per cent of the composition of municipal solid waste is paper, over 10 per cent is glass, 15 per cent is food, about 12 per cent comes from yard and garden materials, 9 per cent is metal, and the remaining 10 or 11 percent is miscellaneous (rubber, plastics, textiles, etc.).

Once the materials are sorted and certain commercial substances removed, communities are faced with the problem of the organic wastes. However, paper products have heat value. In this respect one ton of standard refuse is the equivalent of 1.5 barrels of relatively low-cost fuel oil. Cheap fuel oil is highly polluting and has about 3 per cent sulfur, while refuse is generally low in sulfur content and worth about $4 a ton (fuel oil with as little as three-tenths of 1 per cent sulfur has recently gone up in price to about $12 a barrel in New York City). Today, much of our waste paper is either burned or compressed and buried in landfills, while we continue to hunt for new fuel oil sources.

The United States generates about 360 million tons of refuse each year, with a heat-generating capacity equivalent to one-tenth of our current oil needs (1.5 million barrels in 1973). By 1980, if current growth of waste continues, we will be faced with 540 million tons of refuse, with a heat-generating equivalent of the Alaskan North Slope fields production at that time. This resource in organic wastes can be utilized in our energy-short land within a reasonable period of time.

T. M. Malin, writing in 1972 in *Environmental Science and Technology,* estimated that an 800-bed hospital could reduce its fuel bill $54 a day by properly incinerating its burnable waste, while at the same time saving an estimated $75 a day in the cost of carting it to a dump. The total savings would amount to $47,000 per year. Since space heating is a major consumer of energy, proper burning of waste materials would be a major institutional saving. Western Europe uses its waste for both space heating and power production. The United States has much to learn from other countries in utilizing organic wastes properly.

Failure to utilize such wastes has generated a host of problems in both urban and rural areas. A visitor walking through many high-rise apartment areas of New York and Chicago on a warm moist day is greeted with the smell of dog droppings. The problem grows worse as one approaches the more fashionable sections of these cities. Curbing of animals becomes a major problem when the dog population becomes too large for a given area. Maybe toilets for pets is the answer. This urban problem has a rural counterpart. In certain rural commercial feedlots—and in some parts of the United States the number of feedlots for 100,000 or more cattle is growing—manure accumulates to such an extent that it eventually contaminates streams and air.

In 1971 the journal *Environment* published an article by H. L. Bohn, "A Clean New Gas," in which Bohn suggested that large feedlots could produce enough methane, the main constituent of natural gas, to supply the needs of a city of 30,000. He estimated that combined urban (15 per cent) and agricultural (85 per cent) organic wastes in this country totaled about 1.5 million tons a year. This could potentially result in an annual production of methane equivalent to one and a half times our annual natural gas consumption. This is easier said than done, however, when we remember that our 100 million cattle, 50 million hogs, and 20 million sheep are scattered throughout the country, and it would be nearly impossible to collect all their droppings. Furthermore, the Environmental Protection Agency has determined that this method of disposing of organic wastes requires more fuel than is produced as a by-product. Also, thermal decomposition in an inert atmosphere would create odor problems and could conceivably produce carcinogenic agents.

One of the uneconomical aspects of our technological society is that we have been taught to dilute our sewage by grinding food remains in garbage disposal units. The grinding takes energy, and we then have to use more energy to concentrate the diluted product in our sewage and purify it.

Recycling Centers

Waste disposal and resource recovery go hand in hand. To directly burn organic wastes or turn them into gas for fuel solves part of the problem, but metal and glass materials still litter our streets, fields, and beaches. America has 60 million discarded cars, which number has been increasing by about 2.5 million each year. We are even more aware of the 60 billion single-use beverage containers used each year, better known as throw-aways.

If recycling centers are to remain viable and not to become junkyards, selective profitable materials must be given first priority. Newspapers and cardboard boxes lend themselves more easily to becoming salable recycled paper products than do paper wrappers and other paper containers. Much depends on the proximity of the center, the willingness of citizens to sort materials, and the availability of transportation for hauling materials to the center.

Some glass companies offer as much as $20 a ton, or a penny a pound for glass; some recycling centers do not even accept glass. Aluminum cans are redeemed in some areas for ten cents a pound or $200 a ton, making it a prime reclamation material from an economic standpoint. Kaiser Aluminum's program in California in 1972 was up 80 per cent over the previous year; $535,000 was dispensed to collectors for 118 million cans. At present, the industry boasts that 16 per cent of all aluminum cans produced are being recycled. But even if 50 per cent of all aluminum cans were recycled, their production would require much more energy than if refillable bottles were used instead.

Local recycling centers can be successful only if they are integrated within a complete nationwide system for collecting, reprocessing, and recycling. Recycling centers have opened and closed by the thousands since Earth Day, 1970. While they last they are convenient places to unload both junk and social responsibility if the dynamics of recycling are misunderstood. Volunteer-run recycling centers that cannot collect materials regularly are not going to solve our solid-waste problem. Such centers even have a negative educational effect on the general population when they encourage the wasteful container industry to continue producing cans and nonreturnable bottles at an annual cost of $1.5 billion more than if returnable bottles were used. By advertising recyclability, container manufacturers conceal the high-energy costs of throw-aways. As *Environmental Action* pointed out in March 1974, the Reynolds and Alcoa plants in Massena, New York, consume two-thirds of the energy produced at the St. Lawrence hydroelectric power plant, an amount equal to 10 per cent of the power needs of New York City. At the same time the companies are charged one of the lowest power rates in the country.

Recycling centers must distinguish between materials that are used efficiently and those, like aluminum cans, that are a wasteful exercise of corporation salesmanship. The piles of solid waste mount each year, contributing to the inevitable depletion of the world's resources. Ironically, recycling of traditional waste materials, with the possible exception of scrap iron and steel, is a depressed business. In the mid-1960s there were 15,000 open dumps in the United States. The government decided to close 5,000 of these within a five-year period, but by 1972 the number of open dumps had grown to 17,000 in spite of 3,000 closures, and disposal of solid waste, long a national problem, was rapidly blossoming into a national scandal.

National Strategies

Individual conservation strategies will not work unless they are coordinated with community strategies, but neither will community measures be effective unless they are backed by a strong national conservation program. A rational energy and materials policy must stem from a coordinated—and not a piecemeal—national program. Local patchwork programs place an unreasonable share of the environmental clean-up burden on responsible communities.

National strategies to discourage energy and material consumption should include removal of the depletion allowance, banning of utilities advertising, and, possibly, imposition of fees for disposal of materials. Strategies involving special subsidies for exploration and development of fossil fuel sources help the producers but damage the environment. To the average citizen, some national strategies smack of socialism, governmental pressure, and destruction of individual rights. For the implementation of such strategies an aroused citizenry is necessary. National strategies must possess these characteristics: easy enforcement; fairness to all citizens; implementation in a reasonable time; and effectiveness as a conservation measure. A number of possible national strategies are discussed below.

Transportation Strategies

Nowhere has the need to rise above local and state considerations been more apparent in recent years than in transportation. National transportation strategies affecting automobiles should include:

1. Impose a graduated excise tax on automobiles according to the amount of gasoline consumed. (Thus, cars with energy-consuming conveniences such as power brakes and steering and cars of greater weight and displacement would be made less attractive to consumers.)

2. Require warranties of ten years or more in order to extend the life of the auto, thus cutting in half the amounts of materials required to sustain our auto population.

3. Impose a disposal tax on every car, to be repaid at the time it is junked, thus removing the temptation to abandon cars.

4. Boost the federal tax on gasoline from four cents to nine cents per gallon. Walter W. Heller contends that a tax boost of five cents would induce a cutback of 5 per cent in gasoline consumption and save 130 million barrels of fuel annually.

5. Couple the revenues from this tax (about $5 billion) with the Highway Trust Fund and use the funds for subsidizing alternative methods of transportation and for operating urban subways and buses.

The driver can hardly call these strategies unfair. The cost of maintenance of highways, traffic signals and signs, policing, and landscaping has been borne by *all* citizens, driver and nondriver alike, during the period of vast highway expansion when 200,000 acres of land were being gobbled up each year for the use of auto owners. In fact, user taxes paid by auto drivers cover only one-third of roadway expenses.

An additional strategy would be peacetime rationing of gasoline and other forms of energy. This is never popular but may be the only means of fair distribution during times of emergency.

Energy Tax

Taxing producers of pollutants will not keep dangerous sulfur dioxide emissions from contaminating the air if power companies are able to pay the tax and still make a profit. A far more effective and environmentally sound approach to decreasing energy use and cutting pollution would be to internalize energy costs. One way to do this is by imposing an energy tax, coupled with equity of electric rates and the supply of fuel or energy stamps to the poor. The energy tax could be based on units of fuel used by producer or consumer. A still better approach is to tax according to the theoretical amount of energy that can be obtained from a given material (a tax of so many cents per million British thermal units).

This energy tax should be charged at the source—the oil pump or coal mine—not at the consumer's end. It should be for all forms of energy, both renewable and nonrenewable, since it assumes that both producers and users are culpable for depletion of resources. Such a policy would make the energy tax easier to collect. It would encourage more efficient use of energy sources and would do away with the advantages now enjoyed by industries whose plants are located near cheap hydroelectric power sites and who continue inefficient production techniques.

A further advantage of such a tax is that the money raised could be used to pay for costly externalities, such as cleaning up polluted environments, estimated to be as high as $20 billion a year. The money could also be used to reclaim abandoned strip-mine lands, clean up litter and junk from riverbeds and highways, beautify cities, develop less-polluting energy sources, and furnish the poor with fuel stamps. (It might even be used to provide capital to energy-poor developing nations as compensation for our past and present overuse of cheap energy sources.)

Equity of Electric Rates

Public electric utility rate schedules are generally so structured as to allow for decreasing charges per unit of electricity consumed as total energy use increases. An example of such a rate schedule is the District of Columbia's Public Service Commission Order 5521, giving residential consumption rates for the winter months November through May. There is a minimum charge of two dollars for the first thirty kilowatt hours or fraction thereof. Thereafter the rates decrease as follows:

	Charge per Kilowatt Hour
Next 20 kilowatt hours	3.90 cents
Next 150 kilowatt hours	2.85 cents
Next 200 kilowatt hours	2.55 cents
Next 400 kilowatt hours	1.90 cents
In excess of 800 kilowatt hours	1.35 cents

Such rate schedules include quantity discounts similar to those given wholesalers. In 1966 a little less than a third of the electricity (measured in kilowatt hours) went to residential or domestic users, and two-thirds went to commercial and industrial users.

However, *revenue* from residential and domestic sources came to 41 per cent of the total, and commercial and industrial groups paid only 54 per cent of the total revenues. Even more astounding is a breakdown of these commercial and industrial users into heavy users and light users. Heavy users consumed 45 per cent of the electrical energy but paid only 25 per cent of the total revenue.

A few years ago public utilities were expanding rapidly. Capital costs were high, and supply exceeded demand. As a capital-intensive industry the utilities needed to raise revenue by encouraging demand. The situation is vastly changed today. While still capital-intensive, utilities are no longer in need of rapid expansion. Any claims that further expansion is needed rest on the assumption of an unending growth economy and the artificial need created by the utilities themselves through promotion of energy use, as evidenced by discount structures and advertising. Today, with an energy crisis and massive pollution problems, promotional rate schedules are not only anachronistic but counterproductive.

Two strategies would take energy-use incentive out of rate schedules and at the same time decrease the residential consumer's electric bill, assuming present usage and all other factors remain the same: (1) a *progressive* rate schedule similar to U.S. federal income tax—as electrical use increases so would the per unit price; (2) a single *fixed* rate for each unit of electricity used, regardless of total usage.

The progressive rate schedule is a disincentive scheme, but both schedules would result in considerable savings for the consumer. For instance, in 1970 the average industrial or commercial concern used about 8,000 kilowatt hours (kwhr) of electricity per month at an average rate of 1.17 cents per kwhr, or $94.84 per month. The average residential consumer used only 751.2 kwhr per month at an average rate of 1.47 cents per kwhr, or $11.07 per month. A public utility could charge a flat rate to *all* consumers of 1.3 cents per kwhr and recover the same *total* revenue from all consumers (residential and industrial and commercial). This would increase the average industrial or commercial bill by $10.40 per month and decrease the average residential bill by $1.28, or $15.36 a year.*

---

* Federal Power Commission, *Statistics of Publicly Owned Electric Utilities in the United States,* 1970. These estimates assume the same usage as in 1970.

This unrealized reduction in the average resident's electric bill represents a private subsidy to industry.

Utility companies argue that changing the electric rates will mean higher subway transportation fares and higher consumer prices on many industrial products, but social benefits will be even higher because a decrease in the growing demand for electricity will mean fewer power plants, less pollution, and less use of natural resources. Prices are the single major determinant in energy growth, and to increase the price of electricity to the major users will provide an incentive for them to find ways of cutting down on inefficiencies in their plants.

Encouraging New Technology

Technology is no universal savior, but it can be a faithful servant. A conservation-minded America can find many ways of reducing energy use if new technologies are introduced and old ones modified. Nowhere is this more needed than in the auto industry, where some cars burn four times as much fuel as others. The Environmental Protection Agency performed tests on most domestic and foreign models in 1974 and found differences ranging from Honda Civic's 29.1 miles per gallon in the lightest class to Oldsmobile Toronado's 6.8 miles per gallon in the heaviest class. The average was 11.7 miles per gallon for all the makes tested.

We need a national policy of efficiency in electric production, distribution, and application. The thermal energy generated in making electricity is a wasted resource that pollutes our streams and rivers. America's many cooling towers are stark reminders of our "underdevelopment." The technology exists that can convert this wasted thermal energy for use in heating homes and office buildings, preventing snow accumulation, and desalinating water. The Russians and Swedes already use such thermal emissions for these purposes.

Electric power plants convert only about a third of a fuel's energy into electricity. A promising new-generation technology called magnetohydrodynamics (MHD) utilizes electrical flow in a conducting fluid instead of in the conventional conducting wire. Great savings are predicted in either closed- or open-cycle MHD. The Russians claim to have a test open-cycle MHD in operation. Experts hope to raise electricity generation efficiency to 50 per cent by using such methods. Transmission losses (over 15 per cent of energy generated) could be reduced by both high-voltage transmission lines and the use of cryogenic techniques, although the first method involves certain environmental dangers. Finally, the efficiency of electric motors and appliances could be increased through known technical improvements.

It goes without saying that some of these economies could be achieved with private resources, but applied technology is costly, and an expanded national program of research and development must be also encouraged along the lines of better coal utilization, stack gas cleaning, energy production from refuse and wastes, and development of alternative energy sources.

Recycling Strategies

A practical national recycling policy must include both disincentives to discourage further virgin material production and incentives to encourage an ailing recycling industry. Reduction of use of primary materials through federal restrictions will save on mining energy expenditures, preserve scarce materials, reduce land disturbance, and minimize pollution due to milling and smelting operations. National disincentives should include:

1. Elimination of all depletion allowances and institution of a depletion tax on minerals and fuels in proportion to the proven reserves in this country. Where minerals are in danger of being totally depleted on a global basis, higher taxes must be imposed.

2. Strict controls on mining production and processing that cause environmental pollution, such as strip mining in mountainous terrain, offshore oil drilling, and lead and zinc smelting.

3. Strict emission controls on fossil fuel plants, a moritorium on nuclear power plant construction, and controls on shale oil development in areas where water is scarce.

4. Tighter controls on water pollution from factories and mines and from the transporting of oil and minerals on lakes, streams, and oceans.

5. Removal of favorable freight rates for virgin materials.

Such disincentives will hasten a national materials conservation policy. Environmentally, it is far more costly to produce a ton of steel or aluminum from virgin materials than from recycled products. Such disincentives, when added to an energy tax and equity of electric rates, would go far toward translating these environmental costs into economic costs to users of virgin materials. Today only 18 per cent of aluminum is recycled; environmental regulations would serve to make such secondary materials as scrap aluminum even more attractive. Removing artificial incentives to using virgin materials is paramount. The Environmental Protection Agency says that a 15 per cent depletion allowance on iron ore allows the ore producer to lower his selling price by 13.5 per cent without reducing his profit margin. Removing the allowance would make the price of material produced from virgin iron ore about equal to the price of scrap steel and iron.

Disincentives are especially urgent from a global conservationist viewpoint. Amory Lovins points out in *Development Forum* (June–July 1973) that tales about extraordinarily rich mineral deposits in oceans or other places are cornucopian dreams. "The real issue is not how many atoms of a given metal might be somewhere in that cubic mile of rock, but how they are distributed." He goes on to point out that, although a goodly portion of rich deposits of ores have been found, low-grade ores will demand increased energy costs to extract them—up to a thousandfold increase in some cases. Mining companies use more and more of their income for exploration, and any major new discoveries will buy only a few decades of grace before the world supply is exhausted. At the present rate of consumption, doubling world metal reserves will buy a mere fourteen years.

Recycling and substitution of materials are possible and we are going to have to depend on these to buy the time required to stabilize or reduce the demand for scarce materials. A further disincentive for using virgin materials and for encouraging reclamation would be a disposal fee on materials, equal to the cost it takes to dispose or return a material to productive use. To determine such a fee is not an easy matter, since a fee based on weight would favor lighter competing materials, even when these are quite scarce or involve hidden environmental costs.

Take the case of glass containers versus aluminum cans, two items that have been the center of attention in resource conservation campaigns. The average glass container weighs more than ten times the average aluminum can (one 12-ounce glass bottle weighs approximately 8 ounces, while an aluminum can of the same capacity weighs about 0.67 ounces). Disposal fees per unit weight would mean that glass producers would be charged over ten times the rate of aluminum for an equal number of containers. Needless to say, this would be a boon to aluminum processors, especially since aluminum scrap can be more easily recycled. However, expanded markets for aluminum means far more energy consumed in container production, and it might even encourage more use of virgin bauxite, which is in far shorter supply than the silica needed for making glass.

Disposal fees are a step toward reducing the depletion of natural resources and bringing about an effective recycling program, but that is not enough. Theoretically, nearly all materials are recyclable, but in the present state of resource-recovery technology few materials are economically recyclable. Paper, depending on type and location of the supply, is a borderline recyclable item. Plastics are not amenable to recycling at present. Any scheme to encourage recycling must take these factors into consideration.

Recycling of resources and source reduction must be effected through a combination of complementary strategies, some of which are disincentives and at least one a major incentive: financial assistance in collecting materials that are at present uneconomical for private scrap dealers to collect. It might be necessary to set up a public corporation to collect and reprocess waste materials, a scheme similar to West Germany's program for collecting waste oil. The United States cannot wait until material waste can be collected profitably before it starts a national clean-up of junk. The federal government could lead the way by cleaning up its own junkyards. In Arizona there is an aircraft graveyard containing over 10,000 World War II military pieces. A national clean-up program might begin with this eyesore. It would be labor-intensive and utilize the services of some of our 4 million unemployed. Subsidies for other waste items, partly paid for by disposal fees, would provide an opportunity for small business to break into the fledgling recycling industry.

International Strategies

The often used phrase "Only One Earth" has many interpretations among environmentalists. It means: We have got to make do with our present globe for we have none other. It means: The earth is a single living organism, which, once it becomes sick through pollution, is totally affected. It means: All of us, as a single body of concerned citizens, are duty-bound to care for the earth. It means: Atmosphere, oceans, and general environment recognize no political boundaries.

It means, above all, that piecemeal local, regional, and national strategies and programs are inherently limited unless undertaken in cooperation with the international principles and goals outlined in Chapter 4. Without integration into such global strategies the efforts of the best-intentioned environmentalists will be wasted. Regulations prohibiting ocean dumping by one nation's oil tankers may simply drive customers to the fleet of a nation with more relaxed standards. Regulations against air pollution caused by one plant may encourage the company to move a country willing to exchange air quality for industrialization. Environment recognizes no borders. Neither should regulations. Global programs and strategies for energy and materials conservation should include the following considerations:

1. An expansion of "Earthwatch," the global monitoring program to provide early warning of impending environmental hazards, is urgently needed. Rising levels of harmful chemicals in food in international trade are a growing risk that require monitoring. Oil spills on the oceans are another serious and growing problem. United Nations Environment Program experts have initiated the first steps in a worldwide monitoring project by establishing a list of harmful pollutants.

2. Even if there were general agreement on global hazards, only a few nations have the capability to measure pollutants properly and in such a manner that data can be interchanged with other monitoring stations. Few countries can finance orbiting satellites capable of detecting major oil spills, and many cannot even afford to maintain local monitoring stations. One possible source of revenue to finance an expanded Earthwatch is the imposition of an international energy tax on fossil fuel traffic on the oceans; another is a tax on aircraft crossing the oceans. Revenue from such taxes, collected by various countries at ports and airports, would give the United Nations flexibility to develop its own monitoring and policing programs. When the tax is on consump-

tion of nonrenewable natural resources, it is only proper that part of the revenue be diverted to international research on alternative energy technologies.

3. International groups should lobby in the United Nations for more policing powers. Strict policing of oceans and airspace is necessary. The oceans, covering 70 per cent of the earth's surface, constitute a vast international territory, containing the major source of protein—fish—for a large part of the world's people. Some two-thirds of the world's oil production (approximately 20 billion barrels a year) was transported on these oceans in 1973. With oil shipments increasing each year, oil pollution looms as a threat to the very life of our oceans in the next few decades. Oil dumped on the ocean kills shellfish, contaminates water fowl, and destroys the aquatic plants that provide nutrients for marine life. Most of this pollution is caused by intentional or accidental discharges from tankers and other commercial vessels, from offshore oil fields, and from dumping of waste oil on land and in waterways that eventually finds its way to the oceans. Fuel used on the oceans and other waterways is only 65 per cent utilized; the rest enters the environment. Another source of pollution is the bargeloads of sewage that are dumped daily in ocean areas close to populated areas.

4. An international seabed control authority must be established, or international corporations will carve up the world's seabeds, much as European nations divided up Africa in the 1880s. Such a territorial partitioning would lose for mankind its last major common heritage. An International Seabed Treaty has been proposed and needs to be enforced at the earliest possible moment. Reserves of gas, oil, and minerals worth billions of dollars lie under the oceans, and all peoples, especially those of poorer nations, deserve to share in revenues derived from any development of the seabeds. Revenues should be apportioned according to need. Opposition from the wealthy nations will be massive. A coalition of international nongovernmental organizations will be desperately needed to counter such pressure.

5. An international crusade should be initiated to conserve and share material resources. This should involve collective bargaining under U.N. auspices by consuming nations and organizations representing the raw-material-producing nations, such as the Or-

ganization of Petroleum Exporting Countries (OPEC). "Consuming nations" means not just large-volume consumers but the Third World countries hardest hit by increasing prices of raw materials. The bargaining must include curtailment of production of nonrenewable resources. International social and religious groups could make a major contribution; conserving and sharing is a major tenet of many such groups, and few groups are more apt to energetically espouse the common cause of all men, especially the poor.

Total Material and Energy Savings

A U.S. Geological Survey report issued in the summer of 1973 predicted that a severe shortage of ores and minerals would soon develop unless this country stopped wasting materials. Figure 5 shows how this ravenous appetite for energy will grow in the next few decades, unless something is done to check its growth. This growth in energy consumption is graphically presented in Figure 6 according to type of fuel. A realistic U.S. conservation policy for fuels should hold 1985 energy consumption at the levels forecast for 1975. This could be done with little disruption of our economy, and it calls for no new technology—just use of methods presently known.

The four main areas of energy consumption at the present moment are: transportation, industry, residential and commercial uses, and conversion and transmission losses. In the area of transportation, a series of four coordinated changes could effect major savings in our use of petroleum: smaller cars and other automobile efficiency practices, improved mass transit and bike networks, increased use of railroads for intercity freight, and certain airplane economies.

One of the first national conservation practices adopted during the 1973 Arab oil embargo was cutting back highway speed limits to fifty-five miles per hour. This saved gasoline as well as lives. A far greater saving will accrue by encouraging automotive efficiencies such as use of radial tires and the purchase of compact and economy-sized cars, which received dramatic impetus from the 1973 gasoline shortage. The trend to smaller cars plus a mandatory car tune-up every six months could increase auto gas mileage from

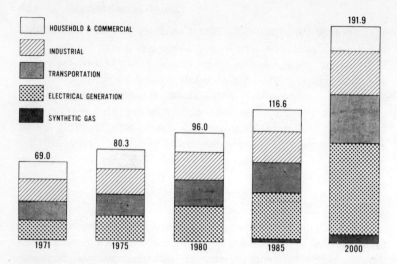

FIGURE 5. U.S. energy consumption by sector, 1971–2000, in quadrillions of British thermal units.

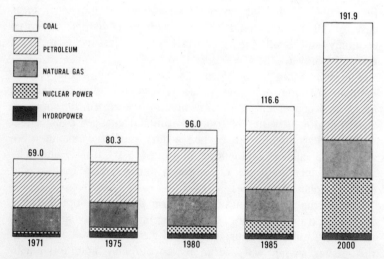

FIGURE 6. U.S. energy consumption by energy source, 1971–2000, in quadrillions of British thermal units.

Source for Figures 5 and 6: Walter G. Dupree, Jr., and James A. West, *United States Energy Through the Year 2000*, Department of the Interior, Washington, D.C., 1972.

about fourteen to twenty-one miles per gallon by 1985 and result in a total energy saving of 4.6 per cent annually.

A second smaller transportation saving could be obtained by shifting about 10 per cent of urban travel from automobiles to bicycles during daylight hours in good weather for trips of five miles or less. Eric Hirst, in an Oak Ridge National Laboratory report, *Energy Use for Bicycling,* calculated that this saving would amount to about 0.3 per cent of U.S. energy consumption. This could be done through a national urban bike network funded by the Highway Trust Fund (since 1973 bike trail subsidies can be so funded) and supplemented by a positive bicycle encouragement program on the part of the Department of Transportation.

Another transportation saving is car-pooling. Here, however, effective national programs have never been fully developed. A much greater saving can be effected through mass transit improvement, since it takes less than half the energy per passenger mile to run a mass transit system than to commute by car. A practical ten-year goal would be to shift intracity passenger travel from the present proportions of 97 per cent auto–3 per cent mass transit to 75 per cent auto–5 per cent bicycle–20 per cent mass transit. This would realize a saving of 0.5 per cent of total U.S. energy expenditures.

A change in long-haul freight rates to favor trains over trucks would be beneficial, since railroads use about 670 British thermal units per ton mile versus 2,800 Btu's for trucks, and 21 per cent of transportation energy is now consumed by trucking. To transfer truck freight to railroad freight or to expand the "piggyback" system for both trucks and passenger cars would mean a saving of 70 per cent. A transfer of only one-fifth of all trucking to railroads would mean a further total energy savings of 0.8 per cent.

Intercity passenger travel could be modified by transferring some of the load to railroads and buses and by reducing inefficient airline travel. Airlines, the third highest user of transportation energy (7.5 per cent), use one-ninth of their fuel on short-haul trips. Transferring of flights of less than 300 miles to other mass transit systems would result in an 0.2 per cent total energy saving. Higher airline load factors, if increased from 55 to 80 per cent, would save 330,000 barrels of fuel per day by 1980 and 450,000 by 1990.

Energy efficiencies in industry would be rapidly improved if the affected industries were made to pay for electricity at the same rates as residential consumers. This incentive alone would induce the industrial sector to reduce consumption by a considerable amount. The oil embargo of 1973–74 made many companies begin to think seriously about conservation. General Electric set a 20 per cent reduction in energy consumption as a goal. Alcoa claimed that its new aluminum process would reduce its energy consumption by 30 per cent. American Telephone and Telegraph hopes to halve its yearly consumption of 12.5 million gallons of diesel fuel. Experts claim that an over-all 20 per cent reduction in energy consumption is reasonable over a ten-year period by reclaiming waste materials, instituting heat-recuperation systems, lowering the intensity of office lighting, and installing more efficient production methods. This would give a major savings in national consumption of about 6.4 per cent.

Resource recovery is another highly potential area of savings, but here little research data is available. If, as suggested earlier, organic wastes were used in municipal power units, one ton of waste would be the equivalent of 1.5 barrels of oil. By 1980 that waste will amount to 540 million tons annually, or 810 million barrels of oil—4.9 per cent of the total energy expended.

Eileen Claussen of the Environmental Protection Agency has calculated energy savings that can accrue if present beer and beverage container systems are converted entirely to refillable glass bottles—244 *trillion* British thermal units at 1972 rates of consumption, allowing ten fillings per bottle, or 0.3 per cent of total U.S. energy consumption.

In the residential and commercial category, the greatest single saving can be effected through better insulation practices. In Table IX space heating is the second largest user of energy (17.9 per cent). By taking into consideration conversion and transmission losses in Table X, space heating amounts to 14.8 per cent of the *total* energy consumption. The 1971 Federal Housing Administration loan requirements for insulation standards should result in 20 per cent savings in fuel or cooling costs. A stricter set of regulations, coupled with a national educational program, could raise this to 40 per cent. By 1985, with an annual construction of 2 million units per year and a net increase of 700,000 units, 38

TABLE IX
Significant End Uses of Energy in the United States

End Use	Per Cent	
Transportation (fuel, excluding lubes and grease)		24.9
Space heating		
Residential	11.0	
Commercial	6.9	
Total		17.9
Process steam (industrial)		16.7
Direct heat (industrial)		11.5
Electric drive (industrial)		7.9
Feedstocks, raw materials		
Commercial	1.6	
Industrial	3.6	
Transportation	0.3	
Total		5.5
Water heating		
Residential	2.9	
Commercial	1.1	
Total		4.0
Air conditioning		
Residential	0.7	
Commercial	1.8	
Total		2.5
Refrigeration		
Residential	1.1	
Commercial	1.1	
Total		2.2
Lighting		
Residential	0.7	
Commercial	0.8	
Total		1.5
Cooking		
Residential	1.1	
Commercial	0.2	
Total		1.3
Electrolytic processes (industrial)		1.2
Other end uses (small appliances, elevators, etc.)		2.9
TOTAL		**100.0%**

SOURCE: *Patterns of Energy Consumption in the United States,* Office of Science and Technology, Washington, D.C., January 1972, pp. 6–7.

per cent of all U.S. homes will be post-1971. A 40 per cent savings on the newer homes and 20 per cent on pre-1971 homes will result in a total energy economy of 4.1 per cent from this sector.

Proper labeling of electrical appliances and appropriate con-

TABLE X
Areas of Major U.S. Potential Energy Savings

Energy Consumption Sector[a] / Area of Savings	Per Cent of Total Savings, 1975–1985[b]
Transportation (25.0%)	
Expanded bike networks	0.3
Smaller cars; radial tires[c]	4.6
Improved mass transit systems	0.5
Increased railroad freight	0.8
Airline short-flight cuts; higher load factors	0.9
Industrial (32.2%)	
More efficient practices	6.4
Power generation from organic wastes	4.9
Using refillable containers	0.3
Residential and commercial (25.1%)	
Better insulation practices	4.1
Cold-water laundering	1.0
More efficient electric appliances	0.9
Conversion and transmission losses (17.4%)	
Better conversion and transmission	5.8
TOTAL	30.5%

[a] Percentages figured from Table IV.

[b] Assuming reductions from about 117 to 80 quadrillion Btu's (Figure 6).

[c] Assuming 55 per cent of transportation sector is automotive, or 13.8 per cent of total energy use.

sumer education could result in small savings for home use of energy. Cooking, lighting, refrigeration, water heating, and small appliances account for about 9 per cent of total energy consumed today in the United States. A 20 per cent improved efficiency on all electric appliances manufactured after 1975 would realize a 0.9 per cent saving, and cold-water laundering would account for an even 1 per cent.

Transmission and conversion savings could be effected by strict Federal Power Commission requirements on both new and existing systems—a possible 5.8 per cent total energy saving.

The series of regulations and practices suggested above could conceivably result in a total energy saving of 30.5 per cent—a roll-back of 1985 energy consumption to 1975 levels.

6 · A Radical Environmentalism

Utopias are needed as goals. Until goals are set, we really don't know whether they can be attained. As a people, Americans are dreamers, and this might be because so many of us come from ancestors who kissed their snug Old World traditional ways goodbye and set out across the Atlantic to make daydreams come true. To strive for Utopia is part of our heritage.

The natural resource crisis has shattered our dreams of unending material wealth and total industrial automation. Technological improvements can't solve all our overconsumption problems. At best, new methods will buy a little time, but they delay our facing up to the needed conservation. Technical innovations are costly—in materials, energy, and money. The solution is not to be found in a technical savior but in a new philosophy that is a healthy mix of traditional thrift, stewardship of the earth and respect for it, a sharing of the world's resources with other peoples, and a strong dose of social welfare.

The question is: How are we going to get middle-class America to return to a life of thrift, to assume its share of stewardship of the earth, and to expand human services and social welfare? At this moment in history, everything seems to be working against our finding an easy answer to this question. We waste enormous quantities of resources; we cut foreign aid and food-relief programs; we gut the Office of Economic Opportunity and other social service organizations.

However, the picture is not totally hopeless. Our country is rapidly becoming aware of world shortages of energy and food; we are becoming conscious of the need for better eating habits; we are becoming environmentally conscious; we were able to voluntarily cut back energy consumption by at least 10 per cent during the winter of 1973–74; and a number of Americans have become interested in a simpler lifestyle. If American environmental utopians are to communicate their hopes to the great majority in our country, they must make it clear that they do not consider government regulations a panacea, that winning legal actions will land us in Utopia, or that the discovery of a new technique for eliminating some pollutant can do the job alone. Government regulations, legal actions, and antipollution techniques are necessary and important, but they are not enough. Utopia will be within our grasp only if we undergo a radical change in our collective lifestyle.

Motherhood, apple pie, and environment. We cherish them all. Most of us like our mothers and most of us like apple pie, and we like a clean environment. We may sometimes be impatient with long-term programs and costly projects, but we are learning to drive slower, to turn down the thermostat, to install more insulation, and to join car pools. We are learning and we are changing.

Corporations, on the other hand, are deeply dissatisfied with strict emission standards and want them relaxed. Electric utilities press for conversion from fuel oil to coal. Auto manufacturers try to postpone compulsory use of emission controls on into the late 1970s. Workers are told that any reduction in consumption will be at their expense. The challenge to utopians is to unite the majority of Americans in a coalition that will be able to respond to such pressures. The road to Utopia was never smooth. Environmentalists, with a far narrower base of support than many consumer groups,* must be given all the support we can muster.

---

* The environmental press is small by commercial standards. Three of the most popular magazines—*Environment, Environmental Action,* and *Not Man Apart*—have a total circulation of about 100,000. A fourth, *Clear Creek,* folded before its first birthday. Some environmental columns receive good circulation in the popular press, but there are virtually no radio or television environmental programs except the Audubon Wildlife Theater. Consumer magazines, by contrast, have widespread popularity. *Consumer Reports* and *Consumer Bulletin* are sold by the millions, and the Consumer Federation of America has millions of members.

How do you make belt-tightening a little more appealing to the American people? Before answering this question, we must first examine the dynamics of individual and social change and see how they apply to our own problems. Only then can we accomplish a radical change in our lifestyle.

Dynamics of Individual and Social Change

"People will have to witness an environmental catastrophe before they'll change."

When we hear or read such apocalyptic statements, we become uneasy. Environmental disasters—famine, destruction of animal or fish populations, widespread contamination by persistent chemicals, nuclear accidents—may be irreversible. They may so damage the fragile structure of this earth that mankind simply could not survive.

Ecological horror stories may or may not move us to action. Quite often the effect is counterproductive—we are paralyzed by fright, or we simply flee to more pleasant surroundings in an effort to forget. A global catastrophe, even if it were reversible, could divert attention from, and actually retard, long-term environmental planning. Fortunately, people can be motivated to change without having to experience the consequences of environmental negligence. You don't have to have cancer to learn of its ravages—or experience a dictatorship to learn the value of freedom.

Americans are practical. An appeal to personal health and safety, love of freedom, quality of life, and respect for the value of the dollar might be of greater effect than the threat of impending ecological disaster.

A national conservation policy must include social as well as individual changes. Environmental degradation and global resource depletion are social issues. Our individual attitudinal changes must be coupled with broadly based social changes. We must seek out and unite with others working for similar goals, such as better health care, better schools and housing, less-congested streets, and adequate mass transit.

We are going to have to expand the U.S. championship of individual rights so as to include social responsibility. The days of enlightened selfishness are over. Social responsibility means an awareness of how an otherwise innocuous act has infringed upon others' rights—an awareness of the fact that a collection of seemingly innocent individual acts can degrade our total environment.

A case in point is the asbestos epidemic plaguing the small town of Manville, New Jersey. Here workers have produced asbestos products for a number of decades for the vast, paternalistic Johns-Manville Corporation. It is not uncommon for a thick swirl of asbestos dust to sweep through these plants and out into the downtown streets. The white dust produces snowstorms in the middle of July in this placid town. As of late 1973 there were 200 to 500 cases of lung cancer in a population of 15,000 and at least 72 cases of mesothelioma, an extremely rare cancer of the lungs and abdominal cavity, which afflicts only one person per million in the general population. The dimensions of the disaster are just now becoming known. Workers and inhabitants who have breathed the dust for thirty to forty years are now developing asbestosis and other cancers associated with asbestos contamination.

In 1967 Dr. Maxwell Borow, a local physician, alerted the medical community to the disaster with a report of seventeen cases of mesothelioma. Two of the victims had never worked in the asbestos plant. Some workers and residents were hostile to outsiders who, they thought, were threatening their job security by talking about a possible epidemic. No doubt the workers had received reasonable pay and fringe benefits, but Johns-Manville refused to reduce asbestos dust levels until forced to do so by federal regulations in 1972. During that same year the corporation spent $25.9 million to build a new corporate headquarters in Denver.

What makes the tragedy worse is that asbestosis, a common asbestos disease, has been known for half a century. As early as 1918, six years before the first report of asbestosis appeared in the medical literature, American and Canadian life insurance companies routinely declined to insure asbestos factory workers. The British Government finally inspected the industry in 1930 and found widespread asbestosis among British workers and required the installation of proper dust suppression and ventilation systems. The first U.S. occupational standard for asbestos, based on a very limited study made in 1938, permitted fifteen times as many parts per million as the standard adopted in 1972 as necessary to preserve health.

The social irresponsibility demonstrated in the Johns-Manville case can be multiplied many times over throughout the United States and other industrialized nations whenever a display of social responsibility for workers' health means reduced profit margins. Industrial blackmail stalks the mining towns of Appalachia and the cotton mills of the Piedmont. The threat of plant closures and loss of jobs keeps local citizens silent, until a major catastrophe occurs. A thousand empty buildings dot the industrial regions of our land, mute testimony to a widespread policy of industrial irresponsibility.

Governmental agencies also bear a social responsibility. The Labor Department has done little to enforce even the mild dust standards now in effect in asbestos plants. The Environmental Protection Agency, after long delay, issued some air standards for asbestos but did not include regulations for construction sites and shipyards, two places where contamination traditionally occurs. The Food and Drug Administration has repeatedly refused to prohibit the use of asbestos filters in processing intravenous drug solutions (not to mention beer and gin), although it is known that the toxic fibers get into such products. This is also believed true of an extensive list of processed foods such as sugar and vinegar. Sadly enough, governmental unresponsiveness simply mirrors that of the rest of our society on such urgent environmental issues.

Within the past decade environmental groups have sought to increase social awareness through action on the legal front. Before 1960 there was a general lack of awareness of environmental hazards. Cases have traditionally been brought to trial only when there was personal physical, psychological, or economic injury, this being a prerequisite to gain the right to a legal standing. Qualifications for the right to sue have been quite strict in American law. When citizen groups are incensed enough to sue, they must still prove they are "sufficiently aggrieved by" or have "sufficient interest in" the pollution problem to gain standing to sue.

The first breakthrough in the right to sue occurred in 1965 in the now famous *Scenic Hudson Preservation Conference* v. *Federal*

Power Commission. Consolidated Edison had asked for a license from the commission to build a power station at Storm King Mountain on the Hudson River. The company plan was opposed by conservationists and three local towns on the grounds that it would spoil the scenic beauty of the area. The U.S. Second Court of Appeals said:

> In order to insure that the Federal Power Commission will adequately protect the public interest in the aesthetic, conservational, and recreational aspects of power development, those who by their activities and conduct have exhibited a special interest in such areas, must be held to be included in the class of "aggrieved" parties.

In 1969 the Sierra Club and other environmental groups opposed the sale of timber from national parks. In the case *Parker* v. *United States* standing was granted on the grounds that the Sierra Club was "advancing the public interest" and had a "special interest in the values which Congress sought to protect."

In 1970 the Sierra Club and other concerned citizens, in *Hudson Valley* v. *Volpe,* opposed a dredge-and-fill operation for the construction of a Hudson River expressway. The plaintiffs contended that the issuance of a permit by the Army Corps of Engineers was beyond the power of the Corps under the Rivers and Harbors Act of 1899. They were granted standing by the Second Circuit Court on the grounds that they had shown that they were responsible representatives of the public interest by evidence of genuine concern through their considerable expense and effort.

However, victories were not to continue indefinitely. In the same year *Sierra Club* v. *Hickle* saw a denial of standing. The Forest Service had taken bids in February 1965 for development of commercial-recreational areas in the Mineral King Valley of Sequoia National Park, and the Department of Interior proposed to permit the construction of an access road into the park by the state of California. The Sierra Club opposed both plans for the access road and for development by Walt Disney Productions. The federal district court granted a preliminary injunction, but on appeal the Ninth Circuit Court reversed the decision, saying the Sierra Club lacked sufficient grounds for standing because it had failed to show sufficient interest in what use the land was put to. (The Administrative Procurement Act requires that a party have a more direct interest than mere concern in order to prove himself "aggrieved" for purposes of judicial review of administrative action.) Furthermore, there was no allegation that the party was injured in fact.

With the general clarification of the right of standing came a host of further actions by public-interest law groups specializing in environmental protection. The Environmental Defense Fund, founded in 1967, has taken on about eighty cases. Sierra Club's Legal Defense Fund, formed in 1971, is quite active, as is the Natural Resources Defense Council, formed in 1970.

Public-interest groups have also utilized new environmental laws that allow class-action suits. The National Environmental Policy Act established a national policy against environmental deterioration that requires Environmental Impact Statements on all federally funded or federally regulated projects that have a potential for environmental damage. The statements are procedural only and do not define what environmental impacts are permissible. Environmental law groups have challenged some of these statements for completeness and have thus delayed projects of a potentially destructive nature. Judge J. Skelly Wright in the historic Calvert Cliffs nuclear power plant decision declared that the Atomic Energy Commission had been making a mockery of the National Environmental Policy Act.

The Clean Air Act of 1970 and the Water Pollution Control Act of 1972 gives the Environmental Protection Agency power to set air pollution and water effluent standards. The Environmental Defense Fund, Florida Defenders of the Environment, and other groups instituted a series of five lawsuits against the construction of the Cross-Florida Barge Canal intended to connect the Atlantic Ocean with the Gulf of Mexico. The charges were that construction would cause irrevocable environmental damage and allegedly violate several federal construction statutes. The court denied the defendant's motion to dismiss and orally granted a preliminary injunction against certain aspects of the construction. Before this order took effect, President Nixon in January 1971 suspended construction on the canal. The Canal Authority of the State of Florida took the President's order to court as a violation of the National Environmental Protection Act, saying that the President does not have power to terminate a project approved by Congress. As of mid-1974, the case was still in the courts.

One of the best known environmental legal actions has been *The Wilderness Society* v. *Morton* (previously Hickle) over the Trans-Alaskan Pipeline.* This proposed multibillion dollar project for extending a 48-inch pipeline for 789 miles through some of the United States' last great wilderness area is considered by many environmentalists to be an abomination. When oil was discovered at the northern Alaska Prudhoe field in 1968, steel pipe was ordered, delivered, and stacked in the hope that the crude oil would flow by 1973. The fields were the largest ever discovered in North America (24 billion barrels of proved oil-in-place and 26 trillion cubic feet of associated dissolved gas).

However, environmentalists doubted that heated oil could be safely transported through frozen tundra and through some of the world's most earthquake-prone regions. Furthermore, the anticipated shipping port of Valdez is situated in one of the world's richest fishing areas, and the outlets from the harbor are quite treacherous to shipping—a fact that conjured up nightmares of oil spills of immense environmental magnitude. An alternative Trans-Canadian Pipeline along the Mackenzie Valley, which could deliver oil to petroleum-starved Midwest areas via Canada instead of the West Coast, would pass through less seismically active regions than the Trans-Alaska Pipeline. U.S. groups pressed for a comprehensive study of the alternative route, to which few Canadians objected, but the impatient oil companies claimed delays were costing $1 million a day.

Legal challenges to the Trans-Alaskan Pipeline were not long in coming. They were based on three claims: (1) the Secretary of the Interior's issuance of a 200-foot right-of-way on both sides of the pipeline violates the Mineral Leasing Act of 1920, which permits a 25-foot right-of-way only; (2) the Forest Service's permit for construction of an oil-tank-farm terminal in the Chugach National Forest violates permit requirements; and (3) the issuance of rights-

---

* Companies that own the pipeline system are: Arco and Sohio (28.08 per cent each), Exxon (25.52 per cent), Mobil (8.86 per cent), Union and Phillips (3.32 per cent each), and Hess (3 per cent).

of-way is in violation of the National Environmental Protection Act because there had been no adequate Environmental Impact Statement and insufficient consideration had been given to the Trans-Canadian Pipeline.

On February 9, 1973, the U.S. Court of Appeals upheld the lower court injunction, ruling only on the violation of the Mineral Leasing Act, and on April 2, 1973, the Supreme Court upheld the U.S. Court of Appeals decision by refusing to hear the case. Thus the matter was thrown into the lap of Congress. There the oil companies mounted one of the largest lobbying campaigns ever waged on an economic issue, overwhelming the rag-tag coalition of environmental lobbyists pressing for further studies of alternative routes. The Senate in a tie vote (broken by the Vice President) voted to allow the construction of the Trans-Alaskan Pipeline. No further legal delays were permissible.

On June 11, 1973, while the pipeline battle was being waged, the Supreme Court upheld the Sierra Club's contention that the Environmental Protection Agency cannot permit a "significant" deterioration of air-quality standards in those areas of the country that already meet or exceed the standards of the 1970 Clean Air Act. This decision was bound to have a major effect on industrial development plans, since 90 per cent of the geographic area of the United States does not suffer from severe air pollution. Environmentalists claim that, if it was not for the Supreme Court's decision, the large electric utility companies would take advantage of vast coal reserves and move into undeveloped regions like the Southwest with the result that the air quality of these unspoiled regions would begin to deteriorate.

Environmental legal groups have won a large number of court battles and have lost very few. The pipeline defeat was by no means typical. It was made possible by the immense lobbying efforts of Big Oil and the panic that seized Congress when faced with the possibility of a major energy crisis. On the whole, the court battles are being won by the environmentalists, but ever since the Arab oil embargo of 1973–74 there has been mounting pressure on Congress to remove or weaken the laws responsible for the environmentalists' victories. The *radical* call now is for citizens who love their country to rally to the defense of its environment and to fight for the preservation of environmental law and order.

Factors in Social Change

Factors needed for social change include: a clear and widely perceived goal, an opportune moment, a basic social structure inviting change, a series of options open to effect the change, a willingness on the part of people to undergo change, and an effective leadership either by a charismatic individual or by some cohesive and inspired group within the system.

The goal is a better environment through less quantitative and more qualitative growth. While the majority might accept this goal, its attainment is not guaranteed. However, the ecological clock ticks away, and current shortages are nudging both capitalist and communist nations in the direction of change.

A free society will theoretically allow options to exist for change, but lack of financial reserves, excessive bureaucracy, the need to earn a livelihood for one's family, political repression, and lack of media accessibility for communication and expression can subtly influence one's choice of options. Social changes require leadership. This may come both through groups (coalitions of citizens or environmental organizations offering collective leadership) or through individuals. Environmental problems, complex and social by nature, require collective direction for solution. No single individual can understand all the engineering, health, social, political, and technical ramifications.

Just as necessary as collective leadership in tackling environmental problems is a series of collective actions for the desired environmental change. These actions, no one of which is totally responsible for the change, can be grouped into two sets: fundamental actions that expose the problem in its entirety and a practical series of actions to fall back on when fundamental actions fail.

Scientific, Informational, and Educational Levels of Action

Fundamental actions—to generate public awareness of particular environmental problems—have been operative for a number of years, even before the first Earth Day of April 1970. However, they were not generally coupled with social-economic-political solutions.

Data gathering and fact finding are necessary, for instance, before we can know whether de-icing with salt results in safer highways as advertised by the Salt Institute, whether this salt enters the waterways and fields causing extensive contamination as alleged by environmentalists, and whether corrosion of autos by salt justifies cost-benefits. We need data about the persistence and use of pesticides, about the type and amount of asbestos found in Lake Superior due to washing of taconite, and on the synergistic effects of various aerosol sprays used in the home in the same hour span.

The salt, pesticide, asbestos, and aerosol industries acknowledge the need for scientific research but quibble over who should do the research and how deeply it should probe. If we were to wait for the industries to conduct the research, it is unlikely that any toxic materials would ever be exposed. Private research centers are traditionally plagued with overly cautious scientific staff, reluctant to reveal dangerous practices. Some do not want to offend their sources of support; others think that making a value judgment as to social dangers is not in their domain and prefer to remain "objective," burying their findings in the technical literature, where it will be read only by a select group of like-minded scientists.

The social effectiveness of basic research quite often depends on the laboratory in which the research is carried out. Scientific experts must make the value judgment that determines whether this or that fact is to be made public, and the medium of communication chosen to reveal the findings determines how public the message becomes. Writing, speaking, and formal discussions are powerful means of exciting environmental action. A series of public debates between environmentalists and industrial producers and other proponents of increased consumption would be one way of presenting both sides of a controversy. So often the corporate advertiser's word cannot be countered because the opposition cannot afford to use the same medium of communication.

A plethora of environmental study courses, degree programs, and projects are available in colleges and universities throughout the country, varying widely in content and purpose. The more creative programs include in-service training with public-interest groups and environmental agencies, so that students can learn about problems through practical experiences. Environmental themes can be used in a variety of educational courses—moviemaking, poetry, painting, general science, economics, and social sciences. Perhaps the most rewarding area for environmental education will be at the grade school level before commercialism has made hardened cynics of the students.

As a generator of environmental action, formal education is quite limited. The biggest handicap of the academic method is that it makes human beings, who after all are a basic part of the environment, into abstract viewers of current events. Certainly, the mental exercise is valuable, but all too often it conveys the impression that problems are more to be studied than remedied, and action is overlooked in favor of leisurely research and writing. Studies, while needed, seldom create revolutions.

Such groups as the ecology centers in various cities and the Environmental Education Group in Los Angeles play an important role in creating an environmental conscience. When citizens don't know where to turn for basic information, these groups fill a major gap. The Professionals in the Public Interest Project in Washington, D.C., tries to link scientific, engineering, and economic talent with citizen groups in need of expert advice on environmental questions. The biggest drawback is lack of adequate funding to give these projects visibility and to staff them with enough personnel to reach citizens throughout the country.

Social, Political, and Legal Actions

There are a number of social, political, and legal actions that can be taken by citizens who feel they are well enough informed to want to do something positive when the fundamental actions described above fail to have any effect. Possible social actions in such cases include:

1. Making the citizen's professional, academic, or civic organization more aware of environmental problems.

2. Actively backing political party candidates who want to deal with environmental issues.

3. Demanding corporate responsibility at annual stockholder meetings.

4. Volunteering to work with public-interest and ecology groups (retirees are of invaluable assistance to such groups).

5. Joining citizen coalitions concerned with problems relating to highways, clean air, and land development.

6. Picketing the homes of local polluters.

Some citizens may not wish to join existing organizations. For instance, an organization may be too conservative to change. If such citizens have organizing ability, they may want to organize a community group for environmental action themselves. The methods of the late Saul Alinski have succeeded in Chicago and other cities in organizing communities around some common issue —better schooling, housing, garbage pickups, police protection, or lower rents. Disorganized communities are the product of environmental degradation, so choose the issue carefully and get trained community organizers to assist if possible. Gather community leaders in a central place and explain the issue clearly. Direct your attack against a visible offender, not an abstract concept or a distant antagonist. Make your actions public as soon as possible and strive for full coverage through the press, radio, and television. A good example of a group that has become quite effective in dealing with inner-city environmental problems is the Urban Environment Conference of Washington, D.C., whose impetus for organization was provided by Senator Philip Hart of Michigan. The Conference is really a coalition of labor, civil rights, and environmental groups that have joined forces in a number of legal and social actions regarding such issues as the removal of lead from gasoline and highway construction projects that threaten the homes of the inner-city poor.

Recent environmental laws allow citizens to bring suits and class actions in the name of a community that has been harmed by polluters. Several organizations have been funded to develop these actions. These include the Environmental Defense Fund and the Natural Resources Defense Council, whose admirable records should be made known to citizens wishing to bring legal action against polluters. Another group of organizations that has sprung up throughout the country recently is the Public Interest Research Groups, started with the aid of Ralph Nader, with state organizations at many major and prominent private liberal arts colleges.

American citizens have access to the offices of their congressmen and should take advantage of it. Lobbying, or influencing legislation, is a time-honored American practice. Until fairly recently the great majority of lobbyists represented special-interest groups that can afford to employ agents to influence legislation. They swarm through the halls of Congress and maintain an elaborate system of information and communication of their own. Citizen lobbies have been established to counter some of this influence.

Citizens can apply political pressure in a number of ways—by:
1. Running for public office or persuading others to do so.
2. Lobbying themselves or supporting public-interest lobby groups.
3. Writing to congressmen. Letters should be short (two to three pages) and to the point. Address a specific issue, outline the problem clearly, and state what remedies should be adopted. Do not duplicate the letters for separate mailings to the same office. Instead, obtain joint signatures and send copies to friends and relatives in other places to send to their own elected officials.
4. Testifying on important environmental issues. When a citizen testifies, send copies of prepared testimony beforehand to the officials conducting the hearing and to the press. When testifying as a group, make sure that material is not covered twice.

Quite often such actions fail to accomplish needed results, or shortage of time may demand more dramatic action. If so, the citizen who is truly free should not feel restrained from approaching (or going beyond) the limits of legal action. This can be done in a number of ways.

Whistle-Blowing. This is the practice of an individual, through his own initiative, bringing to public attention some form of social irresponsibility by an employer or an organization about which he has inside information. Whistle-blowing is risky and often beyond the pale of the law. Legal safeguards against recrimination have not yet been developed to protect the courageous whistle-blower, and such action might lead to expulsion from his job and other forms of recrimination, for example, pressure on his family. Often the notoriety of the action and the courage of the individual can effect more in the way of change than if the whistle-blower had made his cause known through normal legal or political channels. Whistle-blowing is beyond the call of duty, but if a person is willing or anxious to leave his job or is under undue pressure from a superior, he may be tempted to go out with a bang and spill the beans in the process. If you are so tempted, follow these prudent guidelines:

1. Talk the matter over with a confidant. If he does not fully agree and the promptings continue, consult another person you can trust. Each such person should be bound by secrecy.

2. If you decide you must act, and the matter might possibly be resolved by appeal to superiors or to a group of fellow employees or members, try this tactic first.

3. If you must go beyond the walls of the organization, then discuss with friends how this can be done most effectively. And when you blow the whistle, make sure the public hears it. Choose a time when there is usually little news (a weekend perhaps).

4. While you may be willing to take the consequences, don't think the organization that has been guilty of malpractice has more rights than you. Record and document all threats.

The success of the civil rights movement in the middle 1960s was largely due to the passive resistance of men like Martin Luther King. Sit-ins at lunch counters, at schools, and in buses helped to accelerate desegregation. The Gandhi style of nonviolent resistance, right on the border of the legal and often judged illegal by harassed authorities, is highly symbolic and expresses firm faith in a cause and a willingness to sacrifice.

Open Confrontation. Malpractice is sometimes best brought to the public's attention through a process of open confrontation. Daniel Berrigan once said that his many books of poetry and lectures had far less impact on the public than his single symbolic act of pouring blood on draft documents. Open confrontation in environmental causes might take the following forms:

1. Strikes, boycotts, sit-ins at the offices of major polluters.

2. Open disruption of stockholder meetings of irresponsible companies.

3. Lying in front of bulldozers.

4. Signs indicating who a polluter or irresponsible developer is or where he lives.

Ecotage. Ecotage is a still stronger means of effecting environmental change. Bordering on violence, it can be effectively used to counter an act of "silent violence" on the part of industry—to dramatize a violation tacitly condoned by established authorities.

For instance, cutting down billboards that are illegal in the first place would be a form of ecotage. A case that received general press coverage was that of the Chicago "Fox," who took his *nom de guerre* from the Fox River, which he had loved as a child. The so-called Fox slipped into the executive suites of the U.S. Steel Corporation plant that had been polluting waters near Chicago and poured sludge and dead fish on the carpeted floors. He stopped up the smokestacks of air polluters and the sewers of water polluters. Ecoteurs in Florida removed signs from a popular Miami bathing beach and substituted notices that bacteria counts had passed the mark for safe swimming. And on Washington's Birthday in 1974 an Amherst College graduate student, Samuel H. Lovejoy, toppled a 500-foot meteorological tower used for testing sites for nuclear plants.

Putting purple dye in sewer effluents to make wastes highly visible teaches a community a specific lesson and induces environmental discussion. The ecoteur must make sure that the object of his actions is the actual culprit, however, and not an innocent suspect.

Open Disruption. This is a step beyond ecotage—a more blatant form of illegal environmental action that is just short of armed rebellion. Some of the many problems associated with this form of action are:

1. Assembling to discuss open disruption is often termed conspiracy and is thus a crime.

2. The unintentional effects of such disruption may be suppression of desirable legal actions.

3. The participants may be so emotionally motivated that they fail to consider the possible consequences beforehand.

4. It may be difficult to determine whether all legitimate steps have been exhausted before such action is initiated.

Some guiding principles of violent environmental action can be drawn from "just war" theory: (1) The violence should be proportionate to the injury done; (2) human injury should be avoided as much as possible; and (3) other means must first have been exhausted. Consider the case of a strip-mining operation that is moving into a community and has "legal clearance" to destroy one's home and livelihood. The bulldozer is an unjust aggressor. The victim must stop or destroy it before it destroys his home.

All legal steps have been taken but to no avail. The victim can't stop the machine by recourse to law. The only course remaining is to blow up the bulldozer—but at night, when there is no danger of injuring the driver.

If agribusiness is crowding a farmer out of business and all legal means fail, the farmer has no recourse but to make life uncomfortable for the culprit. He knows the lay of the land and is in a good position to conduct guerilla warfare. This case is not as farfetched as it sounds. In the Imperial Valley of California, for example, the major recipients of irrigation benefits from government waterways are large corporations, although the law clearly states that no one with holdings larger than 150 acres should receive the water. Local farmers should therefore feel justified in taking illegal measures against the corporations if their livelihood is threatened.

Many socially desirable changes meet with fierce opposition from the power structure, but opposition is often temporary and self-defeating. Persecution creates martyrs, who, as public witnesses, promote their cause through their suffering. Symbolic action, even the often misunderstood action of going to jail for one's beliefs, may awaken the conscience of the public. We have had no environmental martyrs so far; yet the cause is urgent enough. Symbolic action at the legal borderline or beyond is merely an externalization of the violence that is being done to the world's people through depletion of the world's resources and through pollution. Such violence can be countered in two ways—with passive martyrdom or with acts of direct destruction of property.

Actions speak louder than words.

Toward a Service Economy

The United States in its third hundred years will have to work for a cleaner environment, a reduced use of nonrenewable resources, and a sharing of accumulated wealth with other peoples of the world. The economy will have to change from a materials- and energy-consumption economy to a service-oriented economy, with profit motivation transformed into a desire to improve the quality of life. An enhanced quality of life will involve full employment, equal opportunities for all races, sexes, and age groups, higher-quality education and health services, and opportunities for continued education for all citizens. Actually, the country was moving toward a service-oriented economy in the 1960s, but inflation, lack of leadership, and lethargy reversed some of these gains in the early 1970s.

The biggest obstacle to a service-oriented economy is the massive bite taken each year, even in peacetime, by military expenditures either for present defense systems or in payment for past wars. This bite consumes more than half of the tax dollar, taking with it a large part of the financial resources needed to move the nation in the direction of a service economy.

A vital social environment is peopled by citizens who feel wanted and needed. The U.S. work ethic views labor in terms of production of *goods*. In a healthy social environment work is a meaningful production of *good* by and for people. *Goods* are quantitative, and *good* is qualitative—and that's the basic difference between what we have now and what we should be striving to attain.

Cars, television sets, and airplanes are goods; education, care of the aged, environmental education of the young, continued education of adults, and health services are part of the good. But service by and for the people requires goods also. We need schools, textbooks, medicines, and camp facilities. A service economy does not cease production of goods, but it takes care that what is needed is produced and that what is produced is needed. Fewer cars will be manufactured in a service economy, but they will be of higher quality, longer-lived, safer, and more efficient. The emphasis will be on placing people in service-oriented positions and on viewing their work as *socially* profitable.

There are at least seven service areas where our citizens can find expanded employment opportunities, and perhaps many more if we divert some creative thought to finding them. Characteristics that such opportunities must have include community control (as opposed to bureaucratic or national control), practicality (as to both sources of revenue and operation), less use of natural resources than in the present economy, and utilization of a wide variety of skills and talents.

The service economy will have to balance national and international goals. It will have to balance over-all direction with citi-

A Radical Environmentalism 151

zen control on the local level. A strong movement toward decentralization of power and its return to the people is necessary. Thus, the service areas to be discussed here must have administrative staff and support personnel from and in the community where the work is being done.

Hard-pressed taxpayers are practical. They want to know who is going to pay the bill. I have already mentioned that more than half of the 1974 U.S. budget goes to the military sector; yet a 1973 Brookings Institution report suggested severe cuts in the military part of the budget without any appreciable reduction in the quality of defense. The military departments cut energy consumption drastically during the Arab oil embargo. If we were to cut the military budget in half, it would release $50 billion for needed social services. Other sources of funds would be the energy tax proposed in Chapter 5 (about $4 billion) and income generated from employing 6 million people in the expanded sectors (about $12 billion). The total yield from these three sources would permit employment of 6 million people at pay scales approximately equal to those used to finance Job Corps personnel ($183 million for 17,000 employees in 1973).

The following employment areas require less consumption of natural resources than consumption-oriented industries, and a great variety of skills and talents:

SEVEN AREAS OF SERVICE ECONOMY GROWTH

NEEDS AND REQUIREMENTS	PEOPLE WANTING TO HELP
ADULT EDUCATION	
Adult Needs: All active Americans, not just our youth, should be educated and should continue their education as an integral part of citizenship.	About 2,500,000 instructors and teachers, assuming one instructor for every two classes of 25 students going to school two nights a week.
Service Requirements: 125,000,000 adults need a variety of programs in night courses, in weekend or periodic workshops, or in conferences.	*Skills:* Instructors in languages, remedial and speed reading, crafts, hobbies, dramatics, etc. Professional ongoing instruction by academic people in law, medicine, business, etc.
Energy Needs: Transportation	

152 The Contrasumers

NEEDS AND REQUIREMENTS	PEOPLE WANTING TO HELP
and educational equipment; construction and maintenance costs reduced by use of public school facilities.	

Care for the Aged

Aged Needs: Noninstitutionalized home care for elderly persons lacking it. *Service Requirements:* About 8,000,000 Americans are over 75 years of age. A large number of these need care and companionship. *Energy Needs:* Minimal if existing homes are used; "meals on wheels" require little energy per meal for preparation in bulk amounts.	About 1,000,000 cooks, nurses, and retired people to care for 5–10 persons each (home visits, warm meals, shopping, general care). *Skills:* Retired people in the 65–75 age group willing to share human experiences with others; cooks, nurses, home economists; community supervisors (little job training needed for other employees).

Home Care for the Sick

Considerable hospital care could be given in the home to the satisfaction of all concerned.	About 750,000 nurses, homemakers, and retirees; health supervisors and administrators.

Environmental Clean-up

Needs: A national environmental clean-up program. *Material Requirements:* Collecting equipment. *Energy Requirements:* Part of energy expended in collecting materials would be recovered from salvaged junk.	Laborers, sanitary workers, junk collectors, and youth to remove junk, litter, abandoned houses, and garbage.

Youth Environmental Education

Youth Needs: Environmental education for ages 9–15 for one month a year. Costs would be	About 500,000 (mostly rural) folks to teach urban children to appreciate nature.

NEEDS AND REQUIREMENTS	PEOPLE WANTING TO HELP
comparable to an equal span of public school education. *Materials Requirements:* Lumber for camp construction and upkeep; road materials; recreation and craft materials. *Energy Requirements:* Transportation for urban children to and from camps.	*Skills:* Camp managers, caretakers, cooks, counselors, recreational personnel, carpenters, trail and forestry crews, naturalists, and craftsmen.

VOCATIONAL EDUCATION

Conservation of human resources is closely linked to conservation of raw materials. *Service Requirements:* About 4,000,000 Americans need retraining and job placement. *Materials and Energy Requirements:* More use of present vocational schools and centers, plus new schools in certain regions.	About 400,000 vocational instructors with a wide variety of skills and talents to retrain unemployed and underemployed. *Skills:* Placement officers, counselors, social workers, instructors.

ENVIRONMENTAL POLICING

Local communities need constant checking for environmental deterioration and existence of toxic materials in water, land, and air.	About 100,000 engineers, technicians, scientists, and research staff for a major monitoring program.

The seven service areas would be locally managed and controlled. The services rendered would not be concentrated in certain privileged areas of our country, as are defense industries, but dispersed among regions that have suffered losses of population over the past several decades. The whole thrust of such a service-oriented economy is to decentralize our system and to establish a more steady-state, or balanced, social environment to replace an economy whose existence depends on expenditures of vast quantities of raw materials.

The list of service areas is not meant to be exhaustive. The reader is invited to get into the act of creating new opportunities for the country as we shift from a consumer-based to a more rational culture. The United States in the past has demanded military service from its citizens; it has reneged on demanding social responsibility from industry; and it has slighted its own responsibility to unemployed and underemployed citizens. We must create new service opportunities now to provide employment for those who may lose jobs through the limiting of our economic growth. Only when such actions become national policy will we be on the road to a service-oriented economy.

7 · The Next Hundred Years

To establish and implement a national policy of conservation requires some truly revolutionary changes. It means doing away with some of the undesirable economic and social policies that have infiltrated our American system and introducing new policies more in tune with what our founding fathers intended.

The first major change must be the disestablishment of our national religion: commercialism. Anticommercial-minded heretics are beginning to unite and speak up. They are willing to undergo official displeasure in order to spread a gospel different from that of the marketplace. They gently urge us to change our lifestyle, and they chide us for our hypocrisy in such matters as our generosity to developing nations. From 1960 to 1967, they point out, $9 billion flowed to Latin America from the United States in the form of public and private loans and commercial investments. But the outflow from that region in the form of profits, shipping charges, services, interest on loans, and so on amounted to $19 billion. Our generosity came with a $10 billion price tag.

These anticommercial prophets—call them Contrasumers—say, "Take heed, for time is growing short. If we continue to consume at the present rate, the earth has but a decade or two of life before the whirlwind comes. If we continue to worship our material idols, if we continue to reap profits for the corporations at the expense of the poor, if we continue to neglect the needs of our own people and the other peoples of the world, we will surely die. But if we break up our idols, distribute the profits, and expand our national interests so that they become global interests, then surely we shall be blessed for the next hundred years."

Before a national policy of conservation can be successfully implemented, a number of other changes will have to be made, guaranteeing qualitative improvement of the lives of U.S. citizens through full employment, adequate health care and continuing education for all, environmental education of youth, proper care of the aged and infirm, and job training for the unskilled or semiskilled. The economy will have to be realigned by cutting back production of certain commodities and military expenditures and by initiating a service-oriented economy. A host of strategies will have to be introduced—from improved mass transit systems and equity of electric rates to energy taxes and resource-recovery systems.

The ways in which our citizens desire to qualitatively change their lives should be left to their individual consciences. They should not be required to plod listlessly along on the commercial treadmill when they don't want to. The United States can support a variety of lifestyles and economies. Those desiring to live in an autoless economy should have that right.

While cherished individual freedoms must be preserved, there will have to be some sacrifices of personal freedom of choice for the good of the nation and world. We are not free to commit crimes against our neighbors, and inefficient uses of human and physical resources are crimes. To operate a speedboat without emission and noise controls is a form of indiscipline that weakens the social environment. It denies fishermen the right to peaceful recreation; it denies inhabitants of nearby cities the right to clean water and silence.

Conservation and preservation of resources are not lifestyles in themselves but the means whereby freedom to choose one's lifestyle is ensured. The anticonservationist philosophy of commercialism runs counter to this principle by stifling and repressing the individual's right to choose an alternative lifestyle.

Conservation is not a fad or a temporary belt-tightening until the current fuel shortage is eliminated. Nor is it a symbolic gesture made to secure the good will of poorer nations. It is a global necessity. It exempts no one and requires concerted application of all our technical and economic skills. It affects the very ways we view our world. A lack of faith in the future of our earth, which permits the exploitation of our resources and the pollution of our environment, must not be allowed to harm innocent victims. The belief that there is no future for our earth may be your business, but to profess that belief by destroying my future is intolerable.

The Second American Revolution is coming soon. Americans are an impatient people. They want changes to come fast and with some measure of success. The first American Revolution required seven years of sacrifice, but it gave birth to a constitution that has withstood the ravages of time for more years than any other such document, and out of it came a dream of freedom that has inspired Americans for two centuries. On the threshold of a third century, we—the Contrasumers—are not asking that that dream or that document be changed but that they be made living realities by returning to all citizens the freedoms that have been lost through the economic and social politics now in vogue.

Is that asking so much?

> When our days become dreary with low-hovering clouds and our nights become darker than a thousand midnights, we will know that we are living in a creative turmoil of a genuine civilization struggling to be born.
> —MARTIN LUTHER KING
> *acceptance speech for the Nobel Peace Prize*

LIFESTYLE INDEX

The Lifestyle Index is designed to demonstrate how much energy the individual uses each year and how his standard of living compares with that of the average individual in other countries of the world, as shown in Table XI on pages 178–79. The object is not to pass judgment on the types of productive work and activity the individual engages in but rather to highlight activities carried out at the expense of limited natural resources, especially nonrenewable fossil fuels.

In the final analysis, the quality of life cannot be quantified. The amount of energy used is not a real indication of the quality of one's life, but a better quality of life must include proper and nonwasteful use of energy, and one way of making the individual aware of inefficient use of energy is by calculating his Lifestyle Index. To do that, we must take into account the energy used in productive work, in obtaining nourishment, in traveling, in seeking recreation, and in consuming products of all kinds. Everything in our homes and offices required energy to manufacture—energy to extract the raw materials, to haul the materials to the processing plants, and to bring the finished product to the consumer through a commercial outlet. The end result of these calculations will be to heighten each reader's sense of social responsibility in the efficient use of energy.

In this inventory of energy use, human energy is largely neglected. In most cases physical activity is an alternative to being overweight and having poor muscle tone. When one chooses to go sailing instead of motor boating, to grow vegetables rather than buy them, or to ride a bicycle to work instead of driving a car, it is unlikely that the physical activity required will be so demanding as to necessitate an increased intake of food. The small amount of animal power in use today has also been ignored, but the energy it takes to provide food for people and animals is sizable and is counted in the Index.

Taking an inventory of energy is no easy matter. The Lifestyle Index is the first comprehensive attempt to do this, and many approximations and estimates had to be made. Quantification entails judgments as to what data to include and how to apportion values. Some readers may find a few of these judgments quite unsettling. They may consider it unreasonable to be charged for certain uses of energy that they think do not apply to them. Such a person might ask, "Why should I be allotted 6 per cent of the per capita energy used by the military, even though I object to it as excessive? Why must I be charged for a portion of the energy used by a hospital that I haven't been near during the past year?" But social and governmental services expend energy, too, and it would be misleading to omit such services when comparing our energy expenditures with those of people in other countries.

The Lifestyle Index does not imply that energy is the only factor that should be considered in making decisions in life, nor does it suggest that energy be the only consideration when all other factors are judged insufficient to render a clear choice. A higher quality of life is one in which the variety of choices is maximized, and the waste occurring when these activities are undertaken is minimized.

The Lifestyle Index is divided in six parts: I. Household Energy Expenditures (A. Precise method, B. Approximate method); II. Household Materials and Personal Items; III. Food and Beverages; IV. Leisure Activities; V. Transportation; VI. Social and Collective Services.

The basic unit employed in calculating the Lifestyle Index is the Energy Unit. Each Energy Unit is equivalent to one ten-thousandth of the energy expended by an average U.S. citizen in 1972. (To be more specific, 1 Energy Unit is equivalent to about 10 kilowatt hours or exactly 34,300 British thermal units.) Average values are given [in brackets] for some of the activities listed. These have been computed on the basis of information found in the list of references on pages 180–82. Sources are listed in each such case. In other cases, instructions are given for calculating Energy Unit values with greater precision. Where more than one user is involved, Energy Units are to be summed and divided by the number of users, as indicated.

Part I-A. HOUSEHOLD ENERGY EXPENDITURES
(Precise Method)

The most precise method of calculating household energy expenditures is to convert quantities of electricity and fuel used to Energy Units by applying the following conversion factors. If you have not retained bills for the past year, go directly to Part I-B and use the approximate method. If you do use this precise method, omit Part I-B (except for the section on Residential Building Materials) and omit the section on Preparing and Preserving in Part III (Foods and Beverages) and the section on Electronic Appliances in Part IV (Leisure Activities).

ELECTRICITY

Multiply total kilowatt hours used in the last 12 months by the conversion factor 0.368. _____

NATURAL GAS

Multiply total cubic feet used in the last 12 months by 0.038. _____

FUEL OIL

Multiply number of gallons used in the last 12 months by 4.5. _____

Divide total energy units by number of users in the household and enter total here and on page 177.

Total _____

Part I-B. HOUSEHOLD ENERGY EXPENDITURES
(Approximate Method)

HOME APPLIANCES (*Ref. 1, 2*)

Values listed in brackets are average Energy Units used per item annually. Multiply by the number of items in the home. Appliances used in preparation and preserving of food will be figured in Part III (Foods and Beverages).

162 Lifestyle Index

ELECTRIC APPLIANCES

clock	[6]	_____
floor polisher	[6]	_____
sewing machine	[4]	_____
vacuum cleaner	[17]	_____
air cleaner	[80]	_____
bed covering	[54]	_____
dehumidifier	[128]	_____
heating pad	[4]	_____
humidifier	[60]	_____
germicidal lamp	[52]	_____
hair dryer	[5]	_____
heat lamp (infrared)	[5]	_____
shaver	[0.7]	_____
toothbrush	[0.2]	_____
vibrator	[0.7]	_____
clothes dryer	[365]	_____
iron (hand)	[53]	_____
wash machine (automatic)	[38]	_____
wash machine (nonautomatic)	[28]	_____
water heater (standard)	[1555]	_____
water heater (quick recovery)	[1770]	_____

GAS APPLIANCES

clothes dryer	[277]	_____
water heater	[1170]	_____
	Subtotal	_____

HOME LIGHTING (*Ref. 3, 4*)

Figures are for average annual use. If your household uses more or less, increase or decrease accordingly.

Electric lighting	[268]	_____
Ornamental gas lights	[668]	_____
	Subtotal	_____

COOLING AND VENTILATION (*Ref. 1, 2, 3*)

CENTRAL AIR CONDITIONING (electric)

New England	[755]	_____
Mid-Atlantic	[957]	_____

East North Central	[905]	____
West North Central	[905]	____
South Atlantic	[1510]	____
East South Central	[1560]	____
West South Central	[1710]	____
Mountain	[1058]	____
Pacific	[1210]	____

CENTRAL AIR CONDITIONING (gas)   [1046]   ____

COOLING AND VENTILATING (noncentral)

fan (attic)	[107]	____
fan (circulating)	[16]	____
fan (roll-away)	[51]	____
fan (window)	[58]	____
electric room air conditioner	[316]	____

Subtotal ____

SPACE HEATING (*Ref. 2, 5–8*)

ELECTRICITY, NATURAL GAS, OIL, AND SOLAR HEATING

If you have not used the precise method of Part I-A, the following table will provide an estimate of space heating energy expenditure, depending on location and type of heating.

	Electric	Nat. Gas	Oil	Oil/ Solar
Northeast	6480	5360	6380	—
Mid-Atlantic	5800	4800	5720	2210
East North Central	6030	4900	5940	2290
West North Central	5350	4440	5280	2040
South Atlantic	4460	3700	4400	1700
East South Central	4040	3330	3960	790
West South Central	2900	2400	2860	570
Mountain	4910	4060	4840	1870
Pacific	3800	3140	3740	1440 ____

COAL

Multiply tons of coal used in the last 12 months by 775.   ____

WOOD

Multiply cords used in last 12 months by 620. ____

Subtotal ____

RESIDENTIAL BUILDING MATERIALS AND GROUNDS (*Ref. 2, 8–14*)

Total home construction energy (including mining, processing, fabrication, transportation, and sales) has been allotted over the average lifetime of the home before major alterations are required. A 25-year span has been chosen for amortization. If your residence was built in the last 25 years, add the following for building materials.

single dwelling	[844]	____
2- to 4-unit apartment	[642]	____
5-units or more apartment	[668]	____
public housing	[700]	____

ADDITIONS AND ALTERATIONS

Multiply dollars expended for this purpose during the past year by 1.1. ____

Subtotal ____

LAWN AND GARDEN GASOLINE ENGINES [50] ____

Subtotal ____

Add all subtotals in Part I-B and divide by number of users in the household. Enter total here and on page 177.

Total _____

Part II. HOUSEHOLD MATERIALS AND PERSONAL ITEMS

It requires energy to produce all consumer products from house furnishings to cigarettes. Consumer items such as rugs and soap require energy at every step of processing. In calculating Energy Units, it is necessary to include raw materials, processing, freight, and merchandizing. Handcrafting such household materials as drapes does not result in energy savings, since an electric sewing machine probably uses more energy than if mass production methods were used.

HOUSEHOLD MATERIALS (*Ref. 8, 11, 12*)

The following Energy Units represent annual

expenditures for manufacturing and merchandizing household items. *Operating* energy will be calculated in Part IV for radios, phonographs, and similar items. Energy units are calculated for the individual user. Make allowances if you are a heavy or light user.

furniture	[35]	____
electric appliances	[46]	____
electronics equipment (radio, phonographs, etc.)	[17]	____
pottery, earthenware, china	[5]	____
other household wares (cutlery, glassware, etc.)	[10]	____
rugs and floor coverings	[29]	____
other textile furnishings	[73]	____
tissues, paper towels, and other household paper products	[25]	____
cleaners and soaps	[38]	____
	Subtotal	____

PERSONAL ITEMS (*Ref. 12, 14*)

Adjust, according to whether the individual is a light, normal, or heavy user.

WEARING APPAREL

The larger expenditure in the case of women's apparel is due to larger dollar sales and larger energy expenditures of establishments selling such apparel.

men and boys	[156]	____
women and girls	[204]	____

MISCELLANEOUS

toiletries and beauty aids	[23]	____
health and medical supplies	[58]	____
tobacco products (1.6 packs per day for average cigarette smokers)	[86]	____
shoes, footwear	[27]	____

             other leather products
                (luggage, handbags, gloves)   [6]   ____
             photographic supplies   [6]   ____
             jewelry   [12]   ____
             costume jewelry   [9]   ____
                                              Subtotal ____

        Add all subtotals in Part II and enter total here
        and on page 177.

                                              Total   _____

Part III. FOODS AND BEVERAGES

PRODUCTION AND TRANSPORTATION (*Ref. 14*)

About 12 per cent of America's energy is used to produce, transport, sell, and prepare foods. A large proportion of this is concentrated in meat products. If you do not eat meat, omit the meat category and add extra for vegetables. If you produce your own vegetables by hand and without fertilizer, do not include the 19 Energy Units for vegetables. Energy Units are calculated by apportioning food agricultural energy expenditures according to the dollar basis of the food categories in Reference 12 and making adjustments for imports and exports. Average yearly consumption per capita is given in parentheses.

PRODUCTION

    meat:
        beef (114.8 lb/yr)
        veal (2.2 lb/yr)
        lamb and mutton (3.4 lb/yr)
        pork (73.0 lb/yr)
        chicken (42.9 lb/yr)
        turkey (8.9 lb/yr)   [163]   ____
    dairy products (356.1 lb/yr)   [41]   ____
    eggs (318 eggs/yr)   [11]   ____

Lifestyle Index

vegetables and melons (excluding home gardens; 312 lb/yr)	[20]	———
fruits and nuts (132 lb/yr)	[14]	———
food grain (141 lb/yr)	[7]	———
sugar: refined (103 lb/yr) corn syrup (22 lb/yr)	[9]	———
beverages (coffee, tea, cocoa; 18 lb/yr)	[8]	———
edible vegetable oils and animal fats and oils (53 lb/yr)	[17]	———
fish (11 lb/yr)	[11]	———
	Subtotal	———

TRANSPORTATION OF FOOD TO PROCESSING PLANT AND STORE

Omit if you grow your own produce or buy most of your food at a farm.

	[16]	———
	Subtotal	———

PROCESSING (*Ref. 8, 10–12, 15, 16*)

meat	[28]	———
dairy products	[24]	———
canned and frozen foods	[28]	———
grain products	[16]	———
bakery products	[13]	———
sugar	[21]	———
confectionery	[4]	———
miscellaneous foods	[32]	———
beverages	[24]	———
	Subtotal	———

CONTAINERS (*Ref. 17*)

The following Energy Units are for one 12-ounce container used per day throughout the year. Adjust accordingly.

BEVERAGE CONTAINERS

refillable glass bottles	[24]	_____
one-way glass bottles	[71]	_____
bimetallic cans (steel and aluminum)	[57]	_____
aluminum cans	[96]	_____
distilled beverages or wine	[3]	_____

OTHER FOOD CONTAINERS

paper	[42]	_____
steel (cans)	[17]	_____
glass	[9]	_____
aluminum	[6]	_____
plastic	[7]	_____
	Subtotal	_____

RETAILING AND WHOLESALING ENERGY EXPENDITURES
(*Ref. 12, p. 744; 14*)

Calculated on the basis of grocery sales for various food items.

meat	[46]	_____
dairy products	[18]	_____
fish	[7]	_____
produce	[26]	_____
canned and frozen foods	[30]	_____
grains and bakery products	[17]	_____
sugar and confectionary products	[9]	_____
beverages (nonalcoholic)	[13]	_____
miscellaneous foods	[18]	_____
alcoholic beverages	[31]	_____
	Subtotal	_____

PET FOODS

Multiply the pounds of meat products consumed each week by 50 and the pounds of pet food cereal products by 12 to get the total Energy Units. Then divide this figure by the number of persons in the household who regard the pet as theirs.

Subtotal _____

FOOD PREPARATION AND PRESERVING (*Ref. 1, 2, 4, 18*)

If you calculated home fuels by the precise method in Part I-A, or if all your meals are eaten out, this section should be skipped.

ELECTRIC APPLIANCES

blender	[6]	_____
broiler	[37]	_____
carving knife	[3]	_____
coffee-maker	[39]	_____
deep fryer	[30]	_____
dishwasher	[133]	_____
egg cooker	[5]	_____
freezer (15 cubic feet)	[440]	_____
freezer (frostless, 15 cubic feet)	[648]	_____
frying pan	[68]	_____
hot plate	[33]	_____
mixer	[4]	_____
oven (microwave only)	[70]	_____
range (self-cleaning)	[443]	_____
range (regular)	[432]	_____
refrigerator (12 cubic feet)	[268]	_____
refrigerator (frostless, 12 cubic feet)	[448]	_____
refrigerator-freezer (14 cubic feet)	[418]	_____
refrigerator-freezer (frostless, 14 cubic feet)	[673]	_____
roaster	[75]	_____
sandwich grill	[12]	_____
toaster	[14]	_____
trash compacter	[18]	_____
waffle iron	[8]	_____
waste disposer	[11]	_____

Lifestyle Index

GAS APPLIANCES

outdoor gas grill	[100]	_____
range (apartment)	[350]	_____
range (single unit)	[389]	_____
refrigerator	[509]	_____

Add Energy Units for electric and gas appliances and divide by number of users in household to get subtotal.

Subtotal _____

Add all subtotals in Part III and enter total here and on page 177.

Total _____

Part IV. LEISURE ACTIVITIES

Estimates for leisure activities outside the home vary considerably with distance traveled, means of transportation, equipment used, and frequency of use.

ELECTRONIC APPLIANCES (*Ref. 1, 19*)

Omit if you estimated electricity used by the precise method in Part I-A. Energy expenditures for materials have already been calculated under Household Materials in Part III. Figures are for operation only, computed on the basis of average electric use per item per year (4 hours daily for television, 6 hours weekly for stereo). Adjust accordingly.

radio	[31]	_____
radio/record player	[40]	_____
stereo	[14]	_____
television		
black and white (tube)	[129]	_____
black and white (solid state)	[44]	_____
color (tube)	[243]	_____
color (solid state)	[162]	_____

Subtotal _____

CULTURAL AND LITERARY RECREATION (*Ref. 12, 14*)

LITERARY ACTIVITIES

newspapers	[95]	____
books	[36]	____
periodicals	[15]	____
use of typewriter (materials and operation)	[1]	____
	Subtotal	____

CRAFTS

Activities such as designing, floral arranging, knitting, and sewing require less than 1 Energy Unit and have been omitted. For other activities, add Energy Units according to the following scales, adjusting figures for more intensive, normal, and less intensive use.

[1 to 10 Energy Units]

Painting, drawing, leatherwork, woodcarving, sculpturing, stamp collecting, coin collecting, collecting articles such as bottles. ____

[10 to 100 Energy Units]

Woodworking (with electric lathe), pottery work (with kiln), metalworking (with forge or oven), recording. ____

MISCELLANEOUS ACTIVITIES

visits to amusement parks	[20]	____
motion pictures (8–9 movies per year)	[9]	____
toys	[14]	____
musical instruments	[8]	____
	Subtotal	____

SPORTS (*Ref. 14*)

Outdoor spectator sports require very small amounts of energy (except for night baseball games and the like) and may be omitted. Some participative sports (jogging, hiking, and most field events) require only human energy. Others, such as baseball, football, other ball games, swimming (in unheated pools), surfing, canoeing, skating (on natural ice), fishing (without a motor boat), biking, sledding, and indoor gymnastics (yoga, karate, judo, acrobatics) require less than 1 Energy Unit and may also be omitted. For other participative sports, add Energy Units according to the following scales, adjusting figures for more intensive and less intensive use. Let your conscience be your guide.

[1 to 10 Energy Units]
Indoor basketball, volleyball, wrestling, boxing, squash, handball, outdoor tennis, go-carting, camping, small-boat sailing. _____

[10 to 100 Energy Units]
Tennis (on clay court), skiing, horseback riding, mountain climbing, caving, scuba diving. _____

[More than 100 Energy Units]
Bowling with automatic pins, indoor swimming (private), motor boating, water skiing, snowmobiling, fox hunting, field polo, deep-sea fishing, yachting, airplane flying, dune buggy riding. _____

Subtotal _____

Add all subtotals in Part IV and enter total here and on page 177.

Total _____

Part. V. TRANSPORTATION

All forms of transportation considered below include a factor for associated costs (fuel refining and retailing, vehicle manufacture and maintenance, insurance, etc.) except for highway construction, which is included in Part VI under Public Services. Energy consumed in freighting is included in the consumer product or service and is thus not counted separately here.

Nonbusiness Travel (*Ref. 16, 20–24*)

PRIVATE CAR

This is a major part of your lifestyle energy use and should be calculated as closely as possible. First, determine the number of nonbusiness miles traveled each year by subtracting miles of business travel from annual car mileage. Divide by the average number of passengers. Then multiply this figure by the following number, depending on your car's miles per gallon (mpg).

    8 mpg. Multiply by 0.50
  14 mpg.   "   " 0.29
  20 mpg.   "   " 0.20
  25 mpg.   "   " 0.16   _____

MOTORCYCLE

Divide number of nonbusiness miles each year by average number of riders (if any) and multiply by the following number, depending on miles per gallon.

    50 mpg. Multiply by 0.08
  100 mpg.   "   " 0.04
  160 mpg.   "   " 0.025   _____

GAS, OIL, TIRES, MAINTENANCE, INSURANCE, PARKING

For private car or motorcycle, multiply Energy Units calculated directly above by 0.395. _____

VEHICLE CONSTRUCTION AND RETAILING

This assumes the life of your vehicle is average (about 8 years). Divide the weight of the vehicle by the number of users and multiply this figure by 0.154. _____

OTHER NONBUSINESS TRANSPORTATION

Multiply the number of miles traveled per year by the figures indicated.

train	0.094	_____
highway bus	0.042	_____
urban mass transit	0.14	_____
commercial aircraft	0.30	_____

174 Lifestyle Index

modern cruise liner	0.48	____
yacht	1.4	____
bicycle	0.016	____
	Subtotal	____

RIDING TO WORK (*Ref. 16, 20–27*)

The multiplying factors here are larger than in the preceding section because they include associated costs, such as vehicle maintenance and construction.

PRIVATE CAR

First determine annual mileage by multiplying round trip in miles by workdays per year. Divide by the number of passengers and then multiply this figure by the following number, depending on the car's miles per gallon (mpg).

7 mpg. Multiply by	0.80	
10 mpg. " "	0.56	
14 mpg. " "	0.40	
21 mpg. " "	0.27	
28 mpg. " "	0.20	____

MOTORCYCLE

Divide miles per year by average number of riders (if any) and multiply by the following number, depending on miles per gallon.

50 mpg. Multiply by	0.11	
100 mpg. " "	0.056	
160 mpg. " "	0.036	____

BICYCLE

Multiply miles per year by 0.016. ____

PUBLIC TRANSPORTATION

Multiply miles traveled per year by the following number.

urban mass transit	0.14	____
intercity train	0.094	____
highway bus	0.042	____
	Subtotal	____

Add all subtotals in Part V and enter total here and on page 177.

Total _____

Part VI. SOCIAL AND COLLECTIVE SERVICES

PRIVATE SERVICES (*Ref. 12, 14*)

In addition to the personal uses of energy considered so far, certain social and collective uses must be charged to the ultimate consumer on a per capita basis. We may not attend schools or use hospitals ourselves, but it is necessary to expend energy to keep them available. If you are sure that certain services do not apply to you (for instance, beauty parlors), they may be omitted. Multiply by 2 for heavy use; divide by 2 for light use.

LEGAL	[43]	_____
NONPROFIT (INCLUDING RELIGIOUS)	[56]	_____
PERSONAL		
laundries	[18]	_____
beauty parlors	[8]	_____
barber shops	[3]	_____
photographic services	[2]	_____
shoe repair	[1]	_____
funeral services	[5]	_____
REPAIRS (NONAUTO)	[13]	_____
HOTELS AND LODGINGS	[26]	_____
BUSINESS SERVICES		
advertising, sign painting	[32]	_____
services to buildings	[5]	_____
business and consulting	[12]	_____
credit	[3]	_____
duplicating, mailing, steno	[3]	_____
commercial research and testing	[5]	_____
detective services	[2]	_____

Lifestyle Index

equipment rental [5] ____
trading stamps [3] ____

Subtotal ____

PUBLIC SERVICES (*Ref. 28–30*)

All citizens have access to the following services and facilities. If all of them apply, your subtotal should be 631. The Energy Units listed apply to both construction and operation.

hospitals (public and private) [211] ____
education (public and private) [80] ____
telephone service (3 calls per day) [21] ____
other public utilities [46] ____
highway construction and maintenance [186] ____
conservation and development of resources [11] ____
sewer systems [17] ____
water systems [47] ____
trash collection [12] ____

Subtotal ____

GOVERNMENT SERVICES (*Ref. 8, 11, 12, 16, 20, 31–33*)

Government services are major users of energy. Since all are benefited (or harmed) by these services, the total energy expended must be divided among all citizens. Reliable statistics are available for federal energy expenditures. That is not the case for state and local government services. They are estimated here to be equivalent to the federal government's nonmilitary energy expenditures in comparable areas.

FEDERAL GOVERNMENT

The major user of energy in the federal government is the military, with 603 Energy Units per capita annually. Total federal expenditure, printed as your subtotal, is 709 Energy Units.

Subtotal 709

STATE AND LOCAL GOVERNMENTS
Construction not previously included amounts to 27 Energy Units. Maintenance of fire and police departments, etc., accounts for another 100.

Subtotal   127

POSTAL SERVICES
Per capita energy expenditure for the average person (824 pieces sent or received each year —or about 16 per week) is 31 Energy Units. Adjust accordingly.

Subtotal   _____

Add all subtotals in Part VI and enter total here and in the space below.

Total   _____

Enter below the Energy Unit totals for each of the six parts of the Lifestyle Index as each is completed. Then find your grand total.

Part   *Energy Units*

I. Household Energy Expenditures (A. Precise Method; B. Approximate Method) ...

II. Household Materials and Personal Items ..

III. Foods and Beverages ....

IV. Leisure Activities ....

V. Transportation ....

VI. Social and Collective Services ....

GRAND TOTAL   _____

You should now compare your total annual expenditure of energy with that of the average U.S. citizen (10,000 Energy Units) and with those of citizens of other countries given in Table XI.

TABLE XI
ANNUAL ENERGY UNITS PER CAPITA IN SELECTED COUNTRIES

Country	Units	Country	Units
Afghanistan	23	Germany	4412
Albania	524	Ghana	157
Angola	130	Greece	1240
Argentina	1490	Greenland	3750
Australia	4600	Guatemala	196
Austria	2890	Guinea	85
Bahamas	4285	Haiti	24
Barbados	975	Honduras	183
Bolivia	175	Hong Kong	862
Brazil	435	Iceland	3640
Burma	57	India	157
Burundi	9	Indonesia	106
Cameroon	82	Iran	865
Canada	7870	Ireland	2830
Chad	23	Israel	2245
Chile	1255	Italy	2245
China	473	Ivory Coast	238
Colombia	559	Jamaica	1068
Congo	212	Japan	2755
Costa Rica	378	Jordan	260
Cuba	949	Kenya	145
Czechoslovakia	5590	Khmer Republic	20
Dahomey	30	Kuwait	8610
Denmark	4495	Laos	71
Ecuador	263	Lebanon	709
Egypt	241	Liberia	313
El Salvador	171	Malagasy Republic	62
Ethiopia	34	Mali	21
Finland	3655	Mexico	1072
France	3314	Mozambique	148
Gabon	874	Morocco	171

Nepal	8	Singapore	1320
Netherlands	4325	Spain	1406
Nicaragua	324	Sweden	5140
Niger	21	Switzerland	3015
Nigeria	50	Tanzania	59
Norway	4400	Turkey	436
Pakistan	68	Uganda	61
Panama	662	U.S.S.R.	3825
Paraguay	119	United Kingdom	4650
Peru	519	United States	9500[a]
Philippines	246	Uruguay	775
Poland	3690	Venezuela	2107
Portugal	685	Yemen	11
Puerto Rico	3230	Yugoslavia	1360
Saudi Arabia	813		

WORLD AVERAGES

With United States   1630
Without United States   1167

[a] This represents the per capita U.S. energy expenditure for 1971. The figure for 1972 (latest year on which computations could be based) is 10,000 Energy Units.

SOURCE: *World Energy Supplies,* Statistical Papers, Series J, No. 16, United Nations, New York, 1973 (converted into Energy Unit values by the author).

SOURCES FOR THE LIFESTYLE INDEX

1. Electric Energy Association, New York City, 1973.
2. Robert Griffith, American Gas Association, Arlington, Virginia.
3. John Tansil, "Residential Consumption of Electricity 1950–1970," Oak Ridge National Laboratory, July 1973.
4. "Patterns of Energy Consumption in the United States," Office of Science and Technology, Washington, D.C., 1972, p. 12.
5. Allen Hammond, "Solar Energy: Proposal for a Major Research Program," *Science,* Vol. 179, 1973, p. 1116.
6. *Combustion Handbook,* North American Manufacturing Company, Cleveland, 1965.
7. R. H. Perry, C. H. Chilton, and S. D. Kirkpatrick, *Chemical Engineers' Handbook,* McGraw-Hill, New York, 1963.
8. W. R. Gray and W. M. McCaughey, "1972 Census of Manufacturers: Fuels and Electric Energy Consumed," Washington, D.C., July 1973.
9. W. R. Gray, "1967 Census of Manufacturers: Fuels and Electric Energy Consumed," U.S. Department of Commerce, Washington, D.C., 1971.
10. "Census of Manufacturers, 1967, General Summary," U.S. Department of Commerce, Washington, D.C., 1971, MC67 (1)-1.
11. "Annual Survey of Manufacturers, 1970–71," U.S. Department of Commerce, Washington, D.C., April 1973, M71 (AS)-1.
12. *Statistical Abstract of the United States, 1973,* U.S. Bureau of the Census, Washington, D.C.
13. M. R. Seidel, S. E. Plotkin, and R. O. Reck, "Energy Conservation Strategies," U.S. Environmental Protection Agency, Washington, D.C., July 1973, p. 82.
14. Calculations made independently by the Center for Science in the Public Interest, Washington, D.C., 1974.
15. Eric Hirst, "Energy Use for Food in the United States," Oak Ridge National Laboratory, October 1973.
16. Eric Hirst, "Energy Intensiveness of Passenger and Freight Transport Modes, 1950–1970," Oak Ridge National Laboratory, April 1973.
17. Eileen Claussen, Office of Solid Waste Management Programs, U.S. Environmental Protection Agency, 1974 (personal communication).
18. Bruce Hannon, Center for Advanced Computation, University of Illinois, Urbana, 1974.

19. Anne Paxton, National Association of Broadcasters, Washington, D.C., 1974 (personal communication).
20. E. Hirst and R. Herendeen, "Total Energy Demand for Automobiles," paper presented at Society of Automotive Engineers meeting, Detroit, January 1973.
21. P. N. Yasnowsky and D. S. Colby, "Determining the Effects of Gasoline Price on the Use of Metals in Automobile Manufacture," U.S. Bureau of Mines Report 7871, Washington, D.C., 1974.
22. Eric Hirst, "Transportation Energy Use and Conservation Potential," *Science and Public Affairs,* November 1973, pp. 36–42.
23. Richard Rice, "System Energy and Future Transportation," *Technology Review,* January 1972, pp. 31–37.
24. Eric Hirst, "Direct and Indirect Energy Requirements for Automobiles," Oak Ridge National Laboratory, February 1974.
25. Robert Lindsey, "Flight Cuts and Traffic Gains Offset Airline Fuel Ills in the U.S.," *New York Times,* February 25, 1974.
26. Eric Hirst, "Total Energy Use for Commercial Aviation in the U.S.," in press.
27. Eric Hirst, "Energy Use for Bicycling," Oak Ridge National Laboratory, February 1974.
28. Charles Fritsch, Mendham, New Jersey, 1974 (personal communication).
29. Larry Deister, Office of Public Affairs, U.S. Environmental Protection Agency, 1974 (personal communication).
30. Eric Hirst, "Pollution Control Energy Costs," paper presented at American Society of Mechanical Engineers meeting, Detroit, November 1973.
31. Office of Energy Conservation, "Federal Energy Conservation: An Interim Report," U.S. Department of the Interior, Washington, D.C., October 1973.
32. W. McCaughey and C. F. Taylor, "Shipments of Defense-Oriented Industries 1971," Current Industrial Reports Series, MA-175 (7)-1, U.S. Bureau of the Census, Washington, D.C., 1974.
33. Russ Wylie, U.S. Postal Service, Washington, D.C., 1974 (personal communication).

ADDITIONAL SOURCES

"Census of Mineral Industries," U.S. Bureau of the Census, Washington, D.C., 1971.

Dupree, W. G., and J. A. West, *United States Energy Through the*

Year 2000, U.S. Department of the Interior, Washington, D.C., 1972.

"Energy Statistics," U.S. Department of Transportation, Washington, D.C., 1973.

Hittman Associates, "A Study of Projected Energy Savings by Banning Television Broadcasting During Selected Hours of the Day; Final Report," HIT-568, January 1974.

Hollomon, J. H., *et al.,* "A Study in the Productivity of Servicing Consumer Durable Products and Improvement Alternatives in the Context of Total Life-Cycle Costs," Center for Policy Alternatives and the Charles Stark Draper Laboratory, 1974.

Midwest Research Institute, "Resource and Environmental Profile Analysis of Nine Beverage Container Alternatives," Kansas City, Missouri, 1974.